Lecture Notes in Mechanical Engineering

Lecture Notes in Mechanical Engineering (LNME) publishes the latest developments in Mechanical Engineering—quickly, informally and with high quality. Original research reported in proceedings and post-proceedings represents the core of LNME. Volumes published in LNME embrace all aspects, subfields and new challenges of mechanical engineering. Topics in the series include:

- Engineering Design
- Machinery and Machine Elements
- Mechanical Structures and Stress Analysis
- Automotive Engineering
- Engine Technology
- Aerospace Technology and Astronautics
- Nanotechnology and Microengineering
- Control, Robotics, Mechatronics
- MEMS
- Theoretical and Applied Mechanics
- Dynamical Systems, Control
- Fluid Mechanics
- Engineering Thermodynamics, Heat and Mass Transfer
- Manufacturing
- Precision Engineering, Instrumentation, Measurement
- Materials Engineering
- Tribology and Surface Technology

To submit a proposal or request further information, please contact the Springer Editor of your location:

China: Ms. Ella Zhang at ella.zhang@springer.com
India: Priya Vyas at priya.vyas@springer.com
Rest of Asia, Australia, New Zealand: Swati Meherishi at swati.meherishi@springer.com
All other countries: Dr. Leontina Di Cecco at Leontina.dicecco@springer.com

To submit a proposal for a monograph, please check our Springer Tracts in Mechanical Engineering at http://www.springer.com/series/11693 or contact Leontina.dicecco@springer.com

Indexed by SCOPUS. All books published in the series are submitted for consideration in Web of Science.

More information about this series at http://www.springer.com/series/11236

Yuefan Wei · Shuyun Chng
Editors

Proceedings of the 2nd International Conference on Advanced Surface Enhancement (INCASE 2021)

Innovation Leading to Industrialization

Springer

Editors
Yuefan Wei
Advanced Remanufacturing
and Technology Centre
Singapore, Singapore

Shuyun Chng
Singapore Institute
of Manufacturing Technology
Singapore, Singapore

ISSN 2195-4356 ISSN 2195-4364 (electronic)
Lecture Notes in Mechanical Engineering
ISBN 978-981-16-5762-7 ISBN 978-981-16-5763-4 (eBook)
https://doi.org/10.1007/978-981-16-5763-4

This Springer imprint is published by the registered company Springer Nature Singapore Pte Ltd.
The registered company address is: 152 Beach Road, #21-01/04 Gateway East, Singapore 189721, Singapore

Preface

The past 2020 is a special year, during which the whole world has been surrounded by unprecedented pandemic. The economy, people's life and scientific research have all been greatly impacted, and many international conferences have been postponed or even cancelled. To run the 2nd International Conference on Advanced Surface Enhancement (INCASE 2021), the organizers have experienced painful struggles and considerations, but, in the end, they persisted to hold this conference to keep it as a biennial tradition. This conference brings together academic scientists, engineers and industry partners to exchange and share their achievements and thoughts on advances in surface engineering, surface and sub-surface characterization, surface coatings and modelling and simulation. Deep discussion on the practical challenges and solutions is conducted, and as a result, the dynamic interaction and synergetic collaboration between scientific research and industrial domains is further strengthened. In the light of the COVID-19 new normal, INCASE 2021 is taking place solely virtually on 7–8 Sept 2021, with two-way communication between audience and presenters enabled.

INCASE 2021 conference proceeding helps further broadcast the essential ideas and thoughts of the above topics to the worldwide research fellows. It contains 57 articles and covers four chapters. (1) Surface engineering is a multidisciplinary science and technology that cross-links the process, structure, properties, functions and performance of materials. Understanding the process–structure–performance relationship plays an important role in the sustainable development of advanced manufacturing and remanufacturing tools. The first chapter of "Advances in Surface Engineering" shares the knowledge in the field of surface engineering and its impact on the material properties. (2) Material properties and their degradation are of critical importance in quality inspection to ensure the in-service security and the reliability of engineering components. Their related characterization and analysis have long been an active area of scientific interest. Tremendous efforts have been put into the development of the technologies of material surface integrity inspection. In the second chapter of "Surface and Sub-Surface Characterisation", robust, accurate and high -hroughput techniques, including non-destructive inspection, in-situ monitoring and digitalization, are introduced. (3) Computational simulation is a powerful tool in surface engineering to provide virtual understanding of the changes in structures and properties during surface engineering processes. Material behaviours under real operation parameters and established process–structure–performance correlation through virtual tools are included in the third chapter of "Modelling and Simulation". (4) Surface coatings are predominantly used for protection and decoration of metals and alloys. Coating development is advancing towards being multifunctional and applied to wider applications. The fourth chapter of "Surface Coatings" elaborates the latest development on coating fabrication, functionalization and new applications.

We would like to express our sincere gratitude to all participants for their contributions to INCASE 2021 conference and proceeding and extend our greatest appreciation for the hard work dedicated by the organization staff, scientific committee, industry advisors and symposium chairs. Special thanks to our reviewers for their valuable suggestions and comments. We gratefully acknowledge the financial support from MFN International, Progressive Surface and Tyrida.

Yuefan Wei
Editor-in-Chief

Contents

Advanced in Surface Engineering

Surface Modification of Inconel 718 by Robot-Guided Centrifugal Finishing and Vibratory Finishing

Marius Ohlert[1(✉)], Sebastian Prinz[1], Sebastian Barth[1], and Thomas Bergs[1,2]

[1] Laboratory for Machine Tools and Production Engineering (WZL), RWTH Aachen University, Campus-Boulevard 30, 52074 Aachen, Germany
m.ohlert@wzl.rwth-aachen.de

[2] Fraunhofer Institute for Production Technology IPT, Steinbachstr. 17, 52074 Aachen, Germany

Abstract. The main task of mass finishing processes is the significant reduction of the surface roughness of a workpiece. Since robot-guided centrifugal finishing has a material removal rate up to $Q''_w \leq 500\ \mu m/h$, it is able to reduce the surface roughness in a shorter process time than vibratory finishing. Furthermore, robot-guided centrifugal finishing is a promising manufacturing technology to induce residual compressive stresses even for high-temperature materials, such as Inconel 718. In order to show the potential of robot-guided centrifugal finishing as a surface modification process, experimental tests were carried out. Within the experimental tests, Inconel 718 was machined by vibratory finishing and robot-guided centrifugal finishing. By using robot-guided centrifugal finishing the maximum induced residual compressive stress was 58% higher compared to vibratory finishing. The findings were derived from the difference in terms of transfer of the kinetic energy between the abrasive media and the workpiece in robot-guided centrifugal finishing and vibratory finishing.

Keywords: Robot-guided centrifugal finishing · Residual stress

1 Introduction

In industrial field, mass finishing processes are common to achieve an evenly high surface quality of workpieces with complex surfaces, such as turbine blades [1]. A disadvantage of the different mass finishing processes used so far is the long process time, which results from the low material removal rates [2]. A relatively new mass finishing process with comparatively high metal removal rates, is robot-guided centrifugal finishing. Robot-guided centrifugal finishing is able to achieve a material removal rate of $Q''_w \leq 500\ \mu m/h$ compared to the material removal rate of $Q''_w \leq 10\ \mu m/h$ of vibratory finishing [3]. Due to the handling of the workpiece by a robot, the workpiece can be machined at defined areas [4]. Therefore, robot-guided centrifugal finishing has the potential to machine complex geometries of additively manufactured workpieces economically, because of its possibility of automated machining [5]. In addition to a high surface quality, highly loaded workpieces, such as turbine blades or punches in die and mold manufacturing,

Y. Wei and S. Chng (Eds.): INCASE 2021, LNME, pp. 3–7, 2022.
https://doi.org/10.1007/978-981-16-5763-4_1

often require a surface treatment to enhance the fatigue life or to optimize the tribological properties. Mechanical surface modification processes, such as shot peening, are applied to achieve the required surface treatment [6]. Although robot-guided centrifugal finishing is a mass finishing process, it has the potential to enhance the fatigue life by inducing residual compressive stress in the rim zone [2]. Furthermore, hardening or grain refinement in the rim zone of the workpiece can result from machining by robot-guided centrifugal finishing [7]. However, there is no quantitative comparison between robot-guided centrifugal finishing and vibratory finishing in terms of the resulting residual stress, yet. Therefore, experimental tests regarding residual stresses in the rim zone of the workpiece as function of robot-guided centrifugal finishing and vibratory finishing were carried out.

2 Materials and Methods

In the experimental tests, a casted cuboid workpiece made of the nickel-based alloy Inconel 718 was used. In the beginning, the cuboid workpiece was separated into seven geometrically identical workpieces to achieve the same starting values in terms of surface roughness and residual stress. The dimensions of the testing workpiece were $l \times w \times h = 72 \times 38 \times 48\,\text{mm}^3$. The type of the robot-guided centrifugal finishing system was Surf-Finisher RSF 700 by the company Rösler and the type of the vibratory finishing machine was Rotary vibrator R 125 by the company Rösler, shown in Fig. 1.

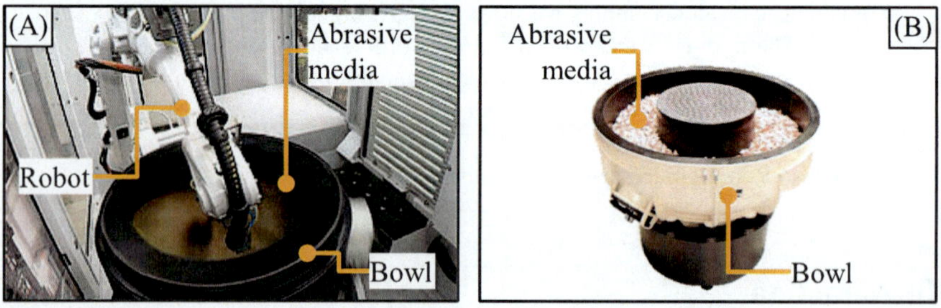

Fig. 1. Illustration of the used robot-guided centrifugal finishing system Surf-Finisher RSF 700 (A) and vibratory finishing system Rotary vibrator R 125 (B)

The process parameters for robot-guided centrifugal finishing are shown Table 1. The process parameters were chosen based on industrial practice to induce residual compressive stress.

The process parameters for vibratory finishing are shown in Table 2. The process parameters were chosen based on the findings of Lachenmaier to induce residual compressive stress in the rim zone of the workpiece [8]. For the robot-guided centrifugal finishing, the process time was not as long as for vibratory finishing. Both processes were carried out with the compound ZF113 of the company Rösler.

Table 1. Overview about the process parameters for robot-guided centrifugal finishing

Robot-guided centrifugal finishing (RCF)	
Rotational speed of the bowl n_{RCF}	85 min^{-1}
Radial position r_{RCF}	250 mm
Immersion depth d_{RCF}	150 mm
Tilt angle δ_{RCF}	20°
Abrasive media	Ceramic (RCP 09/09 S, Rösler)
Process time t_{RCF}	30 min

Table 2. Overview about the process parameters for vibratory finishing

Vibratory finishing (VF)	
Rotational speed of the motor n_{VF}	1,650 min^{-1}
Upper imbalance mass $m_{u,VF}$	4.65 kg
Lower imbalance mass $m_{l,VF}$	3.85 kg
Imbalance angle α_{VF}	90°
Abrasive media	Ceramic (RPM 25K, Rösler)
Process time t_{VF}	120 min

3 Results and Discussion

The experimental results of the residual stress in the initial state (INI) and after machining by robot-guided centrifugal finishing (RCF) and vibratory finishing (VF) are shown in Fig. 2. The tests were carried out three times for each technology.

It can be seen that robot-guided centrifugal finishing in particular increases the residual compressive stress in the rim zone of the workpiece. The maximum induced residual compressive stress was 58% higher with robot-guided centrifugal finishing at approximately $\sigma_{RCF,max} = -600$ MPa than with vibratory finishing at approximately $\sigma_{VF,max} = -380$ MPa. Furthermore, the maximum residual compressive stress induced by robot-guided centrifugal finishing was present at a much higher depth ($d_{m,RCF} = 200\ \mu m$) in the rim zone of the workpiece than the maximum residual compressive stress induced by vibratory finishing ($d_{m,VF} = 80\ \mu m$).

It can be assumed, that these findings resulted from the occurring kinetic energy, which is transferred within the contact between the abrasive media and the workpiece. This kinetic energy is depending on the relative velocity in mass finishing processes. Since the relative velocity is about 300% higher in robot-guided centrifugal finishing [3] than in vibratory finishing [8], the kinetic energy is also higher.

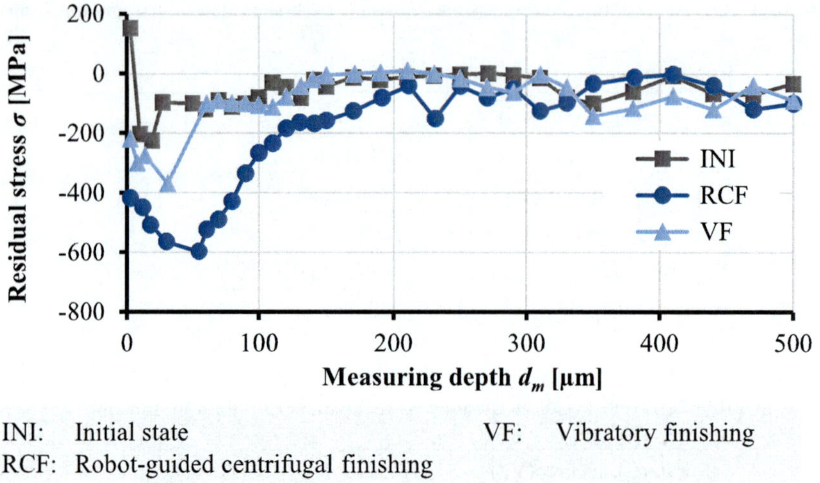

Fig. 2. Illustration of the residual stress in the rim zone of the workpiece depending on the measuring depth before machining (initial sate) and after machining by robot-guided centrifugal finishing and vibratory finishing

4 Conclusion

In this paper, Inconel 718 was machined by robot-guided centrifugal finishing and vibratory finishing. Robot-guided centrifugal finishing achieved 58% higher residual compressive stress and 150% higher depth in terms of residual compressive stress than vibratory finishing.

Acknowledgements. All authors acknowledge the financial support provided by the German Research Foundation (DFG) in the project BE 2542/31-1 "Gestaltungsmodell zur wissensbasierten Einstellung der Randzonen- und Oberflächeneigenschaften additiv gefertigter Bauteile für das geführte Fliehkraftgleitschleifen".

References

1. Klocke, F.: Manufacturing Processes 2 – Grinding, Honing Lapping. Springer, Heidelberg (2009)
2. Schulze, V., Gibmeier, J., Kacaras, A.: Qualification of the stream finishing process for surface modification. CIRP Ann. **66**, 523–526 (2017)
3. Itoh, S., Ho, J., Turangan, C., Wan, S.: In situ measurement of granular pressure and velocity on component surfaces in stream finishing. In: Itoh, S., Shukla, S. (eds.) INCASE 2019. LNME, pp. 226–233. Springer, Singapore (2020). https://doi.org/10.1007/978-981-15-0054-1_23
4. Brocker, R., Klocke, F.: Das Fertigungsverfahren Gleitschleifen – Potentiale in der Oberflächenbearbeitung. In: wt Werkstattstechnik online, vol. 6, pp. 385–389 (2011)
5. Turangan, C.K., Wan Yee Ming, S., Itoh, S., Ho, J.: Development of rheology and computational flow model for robotized external finishing on additively manufactured components. In: Itoh, S., Shukla, S. (eds.) INCASE 2019. LNME, pp. 234–243. Springer, Singapore (2020). https://doi.org/10.1007/978-981-15-0054-1_24

6. Schulze, V., Bleicher, F., Groche, P., Guo, Y.B., Pyun, Y.S.: Surface modification by machine hammer peening and burnishing. CIRP Ann. **65**, 809–832 (2016)
7. Kacaras, A., Gibmeier, J., Zanger, F., Schulze, V.: Influence of rotational speed on surface states after stream finishing. Procedia CIRP **71**, 221–226 (2018)
8. Lachenmaier, M.: Beschleunigungsbasierte Prozessauslegung des ungeführten Vibrationsgleitschleifprozesses. Dissertation RWTH Aachen University (2020)

Check for updates

Advanced Surface Enhancement of a High Strength Aluminum Alloy Through Friction Stir Processing

Ali Hosseinzadeh[1,2](✉), Amin Radi[1,2], and Guney Guven Yapici[1,2]

[1] Department of Mechanical Engineering, Ozyegin University, Istanbul, Turkey
{ali.hosseinzadeh,amin.radi}@ozu.edu.tr,
guven.yapici@ozyegin.edu.tr

[2] Mechanics and Manufacturing of Functional and Structural Materials Laboratory (MEMFIS), Ozyegin University, Istanbul, Turkey

Abstract. The current study investigates the effect of severe plastic deformation (SPD) on the mechanical behavior of a commercial aluminum alloy (Al7075). In-process cooled friction stir processing (FSP) was used to enhance the mechanical properties of annealed Al7075. Substantial enhancements in the microhardness, strength, and strain at failure were achieved through the application of this SPD technique without further thermal treatments.

Keywords: Aluminum alloy · Severe plastic deformation · Friction stir processing · Mechanical behavior

1 Introduction

Aluminum and its alloys are the main preferred light materials for aerospace and automobile industries as well as marine applications due to their high strength to weight ratio, good formability, high thermal conductivity, and high toughness [1, 2]. Among other aluminum alloys, 7xxx series and especially Al7075 has desirable properties such as high strength and high corrosion resistance. Furthermore, favorable precipitation hardening features of this alloy is an important factor in utilizing it for various industries [3]. Employment of thermo-mechanical processing can help develop crystallographic textures and affect the yield strength (YS), ultimate tensile strength (UTS), and elongation to failure [4, 5].

Friction stir processing (FSP) is among the most preferred SPD methods for generating ultrafine-grained (UFG) microstructures on the surface [6, 7]. According to the Hall-Petch equation, grain size dictates the strength of polycrystalline materials [8]. Fabrication of UFG materials improves strength but decreases ductility and thus necessitates further post-SPD treatments to recover a decent ductility level. FSP processing with optimized parameters has the capability of eliminating the stress-strain trade-off [9] as well as modifying the high temperature flow response [10]. Although FSP has been employed in several attempts for manufacturing of metal matrix composites and

Y. Wei and S. Chng (Eds.): INCASE 2021, LNME, pp. 8–12, 2022.
https://doi.org/10.1007/978-981-16-5763-4_2

improving the mechanical behavior of aluminum alloys via surface enhancement [11], the potential of obtaining tough microstructures in Al7075 without post-FSP treatments is yet to be investigated. In particular, the current study aims to determine the effect of in-process cooled FSP on the hardness and tensile properties of Al7075. The significance of in-process cooling for reaching at optimized mechanical response is shown.

2 Experimental Procedure

The as-received material used in this work was a 7xxx series aluminum alloy in annealed condition (Al7075-O) with the chemical composition listed in Table 1. Slabs with dimensions of 20 × 100 × 3 mm was used for the application of FSP.

Table 1. Chemical composition (wt.%) of Al7075 aluminum alloy.

Element	Zn	Mg	Cu	Fe	Cr	Si	Mn	Ti	Al
Wt.%	5.84	2.59	1.53	0.21	0.19	0.11	0.05	0.03	Balance

In-process cooled FSP was performed on the samples to observe the influence of SPD on the mechanical properties. H13 steel tool with the dimensions mentioned in Table 2 was used. After a series of preliminary trials, optimum parameters were selected for FSP under high-pressure (6 bar) air cooling directed on the sample (Table 3). Schematic of the FSP process is demonstrated in Fig. 1.

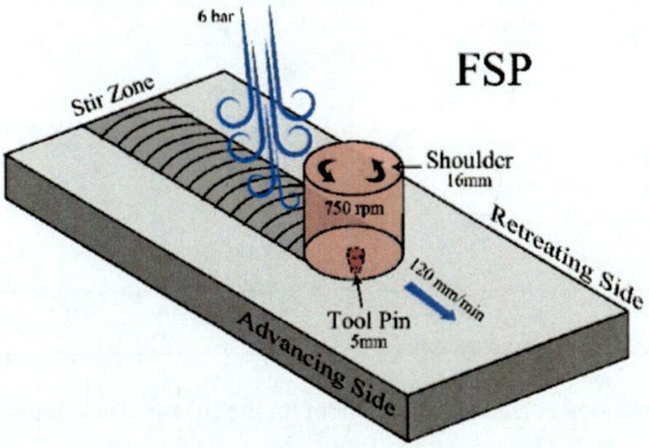

Fig. 1. Schematic of the friction stir processing applied for surface enhancement.

Table 2. Specification of tool profile and dimensions.

Profile of pin	Pin diameter (mm)	Length of pin (mm)	Shoulder diameter (mm)
Cylindrical	5	2.75	16

All the samples were cut via electro-discharge machining (EDM) then ground and mechanically polished to get rid of the adverse effects of surface defects and residual stress. Microhardness testing was conducted with the application of 1000 gf for 15 s. Dog-bone shaped tensile specimens with the 15 mm gage length were tested in the quasi-static regime at ambient temperature and at a strain rate of 10^{-3} s^{-1}.

Table 3. Optimized parameters used during FSP.

Plunge depth	Rotational speed (rpm)	Traverse speed (mm/min)	Cooling air pressure (bar)
Cylindrical	750	120	6

3 Results and Discussion

Figure 2a demonstrates the microhardness tests with an interval distance of 0.5 mm were applied for the FSPed sample to observe the microhardness variation throughout the different processing regions of stir zone (SZ), thermo-mechanically affected zone (TMAZ), and heat-affected zone (HAZ). The microhardness of as-received Al7075 increased from 72 HV to 162 HV at the SZ after FSP. This can be attributed to the extensive grain refinement and stirring effect of the pin at the SZ during FSP [1, 10, 12].

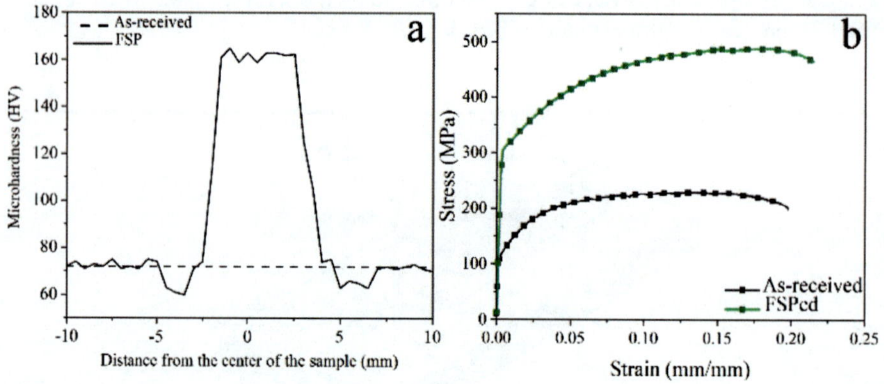

Fig. 2. Effect of FSP induced surface enhancement on the (a) microhardness and, (b) stress-strain response of Al7075.

The tensile response of Al7075 for FSPed and as-received samples are illustrated in Fig. 2b. It is obvious that SPD via FSP improved the YS significantly. This enhancement is related to the grain refinement combined with possible formation of precipitates throughout the FSP [13]. Favorable UTS and strain to failure levels were achieved after in-process cooled FSP with the optimum parameters listed in Table 3. The YS and UTS of the FSPed sample increased to 320 MPa and 480 MPa with uniform strain hardening, respectively. This enhancement may be linked with the grain refinement and recrystallization during the intense stirring exerted by the pin of the FSP tool [1, 10] and the increased volume fraction of nano-sized η' precipitates [13]. Furthermore, employment of the high-pressure in-process air cooling decreases the effect of heat input and subsequently prevents noticeable grain growth at the SZ throughout FSP [13, 14].

Based on the results, the main advantage of in-process cooled FSP is the improvement of the mechanical properties in one step without further post-SPD treatments. Also, strain hardening capability was restored such that the elongation to failure level of the FSPed sample exceeded that of the as-received Al7075. It may be asserted that the in-process cooling has an instrumental effect in balancing out the high heat input to avoid excessive recovery while inducing the formation of coherent precipitates. As such, the proposed technique is deemed to be viable for one-step surface enhancement.

Further characterization on the mechanical response of the enhanced surface is under way.

4 Conclusion

This study investigated the effect of SPD via in-process cooled FSP on the hardness and tensile behavior of Al7075 in annealed condition. The highest microhardness of 162 HV was achieved while the YS and the UTS improved up to 320 MPa and 480 MPa, respectively without diminishing the ductility of the as-received condition. The in-process cooled FSP was demonstrated to be a viable surface enhancement technique for the optimization of mechanical properties without the application of post-SPD thermal treatments.

References

1. Hosseinzadeh, A., Yapici, G.G.: High temperature characteristics of Al2024/SiC metal matrix composite fabricated by friction stir processing. Mater. Sci. Eng. A **731**, 487–494 (2018)
2. Rana, R., Purohit, R., Das, S.: Reviews on the influences of alloying elements on the microstructure and mechanical properties of aluminum alloys and aluminum alloy composites. Int. J. Sci. Res. Publ. **2**(6), 1–7 (2012)
3. Panigrahi, S.K., Jayaganthan, R.: Effect of ageing on microstructure and mechanical properties of bulk, cryorolled, and room temperature rolled Al 7075 alloy. J. Alloy. Compd. **509**(40), 9609–9616 (2011)
4. Jayaganthan, R., Brokmeier, H.-G., Schwebke, B., Panigrahi, S.: Microstructure and texture evolution in cryorolled Al 7075 alloy. J. Alloy. Compd **496**(1–2), 183–188 (2010)
5. Panigrahi, S.K., Jayaganthan, R.: A comparative study on mechanical properties of Al 7075 alloy processed by rolling at cryogenic temperature and room temperature. Mat. Sci. Forum **584**, 734–740 (2008)
6. Su, J.-Q., Nelson, T.W., Sterling, C.J.: Microstructure evolution during FSW/FSP of high strength aluminum alloys. Mater. Sci. Eng. A **405**(1–2), 277–286 (2005)
7. Yapici, G.G., Ibrahim, I.J.: On the fatigue and fracture behavior of keyhole-free friction stir spot welded joints in an aluminum alloy. J. Market. Res. **11**, 40–49 (2021)
8. Jawad, D.H., Hosseinzadeh, A., Yapici, G.G.: Effect of layer architecture on the mechanical behavior of accumulative roll bonded interstitial free steel/aluminum composites. Mat. Sci. Eng. A **818**, 141387 (2021)

9. Yapici, G.G.: Simultaneous improvement in strength and ductility of severely deformed niobium alloy. Mat. Lett. **279**, 128443 (2020)
10. Hosseinzadeh, A., Yapici, G.G.: On the high-temperature flow response of friction stir processed magnesium metal matrix composites. J. Eng. Mat. Technol. **143**(1), 011005 (2021)
11. Patel, S.K., Singh, V.P., Roy, B.S., Kuriachen, B.: Recent research progresses in Al-7075 based in-situ surface composite fabrication through friction stir processing: a review. Mat. Sci, Eng. B **262**, 114708 (2020)
12. Ma, Z.: Friction stir processing technology: a review. Metall. Mater. Trans. A. **39**(3), 642–658 (2008)
13. Humphreys, F.J., Prangnell, P.B., Priestner, R.: Fine-grained alloys by thermomechanical processing. Curr. Opin. Solid State Mater. Sci. **5**(1), 15–21 (2001)
14. Kumar, A., Godasu, A.K., Pal, K., Mula, S.: Effects of in-process cryocooling on metallurgical and mechanical properties of friction stir processed Al7075 alloy. Mater. Charact. **144**, 440–447 (2018)

Vibratory Finishing of Laser Powder Bed Fused Stainless Steel 316 Internal Cooling Channels

Boon Loong Toh[1,2(✉)], Sharan Kumar Gopasetty[1,2], Arun Prasanth Nagalingam[1,2,3], Joselito Yam II Alcaraz[1,2], Zhang Jing[1,2], Swee Hock Yeo[1,2], and Abhay Gopinath[3]

[1] School of Mechanical and Aerospace Engineering, Nanyang Technological University, 50 Nanyang Avenue, Singapore 639798, Singapore
boonloong.toh@ntu.edu.sg
[2] Rolls-Royce@NTU Corporate Lab, N3.1-B2a-01, 50 Nanyang Avenue, Singapore 639798, Singapore
[3] Rolls-Royce Singapore Pte Ltd, Singapore, Singapore

Abstract. The Laser-powder bed fusion process (L-PBF) offers the advantage to manufacture tools and dies with cooling channels precisely following the component contour. The L-PBF manufactured cooling channels have poor surface finish quality. We enhance the internal surface quality of cooling channels—to reduce the risk of ineffective cooling—through the vibratory finishing process. Surface finishing reduces the cooling time drastically to increase production efficiency. We improved the surface finish quality of linear and non-linear L-PBF stainless steel 316 cooling channels. To improve the media flow inside the cooling channels during the vibratory finishing process, we used 98.4% mass concentration of satellite media, 0.738% mass concentration of water, and a 0.837% mass concentration of 10 μm sized SiC abrasives. We polished the channels for 5 h. Additionally, we conducted a discrete element method (DEM) simulation to understand media flow in the differently shaped channels when subjected to vibratory finishing. Lastly, we discussed the changes in channel surface characteristics and suggested methods to surface finishing the more complex internal channels through vibratory finishing.

Keywords: Vibratory media finishing · Additive manufacturing · Cooling channels · Surface roughness

1 Introduction

Conformal cooling systems are used in a wide range of industries such as tooling, automotive, aerospace, medical and electronics [1–3]. As opposed to conformal channels, conventional straight-drilled channels have poorer cooling performance [4, 5]. Conformal cooling channels wrap closely with the channel geometry, enabling uniform and hence, faster cooling of the part [4]. Faster cooling of the part leads to a significant reduction in cost since part cooling comprises nearly 50% of a typical injection molding cycle time [6].

Additive manufacturing (AM) is commonly used for producing conformal cooling channels due to the channels' complex geometry. However, AM channels without post

Y. Wei and S. Chng (Eds.): INCASE 2021, LNME, pp. 13–16, 2022.
https://doi.org/10.1007/978-981-16-5763-4_3

surface finishing processes suffer from high surface roughness up to 20 μm [7, 8]. Reduction and control of surface roughness in cooling channels are required for improved heat transfer performance.

This study explores the use of vibratory finishing (VF) for the surface finishing of AM printed internal channels. Understanding media flow and surface finishing capabilities of VF will be useful in the optimization of future VF experiments.

2 Vibratory Finishing Methodology

2.1 Workpiece and Measurement Methodology

The workpieces are additively manufactured using selective laser melting. The stainless steel (SS316) workpieces designed are shown in Fig. 1. The line surface roughness is measured using Mitutoyo Surftest SJ-400 and the areas are shown in Fig. 1. Roughness measurements are performed every hour. The diameter of the channels are 10 mm, since cooling channels typically have a diameter that is 4 to 14 mm [4].

2.2 Media and Equipment

The vibratory finishing of the channels was performed using a mixture of polishing media and abrasive particles, which were speculated to enhance media flow and material removal capabilities. The media used is a stainless-steel satellite media, measuring 7 mm in diameter and 5 mm in height. The abrasive particles used are 10 μm Silicon Carbide (SiC). SiC particles and water are gradually added to produce a uniform coating of abrasives on the polishing media. The final percentage weight distribution of the paste mixture is 98.4% of steel media, 0.837% of Silicon Carbide abrasives, and 0.738% of water. The vibratory trough is a Walther Trowal TFM 58/32, operated at the frequency of 1200 rpm. The workpieces are held one at a time by a fixture attached to the vibratory trough. The workpiece is oriented such that media flow is parallel to the centerline of the channel entrance.

3 Results and Discussion

3.1 Experimental Results

Figure 2 shows the surface roughness parameters at the straight channel, right-angled channel, and obtuse-angled channel, for every 1-h intervals. The Ra and Rz values are recorded as the average of 5 measurements. In general, the straight and obtuse-angled section shows saturation of the roughness parameters at 3 h of polishing. Overall, the best finishing results is seen at the straight channel, having a 39.5% and 38.9% reduction in the Ra and Rz respectively. The poorest finishing results are seen in the right-angled channel, having a 12.9% and 14.6% reduction in the Ra and Rz respectively.

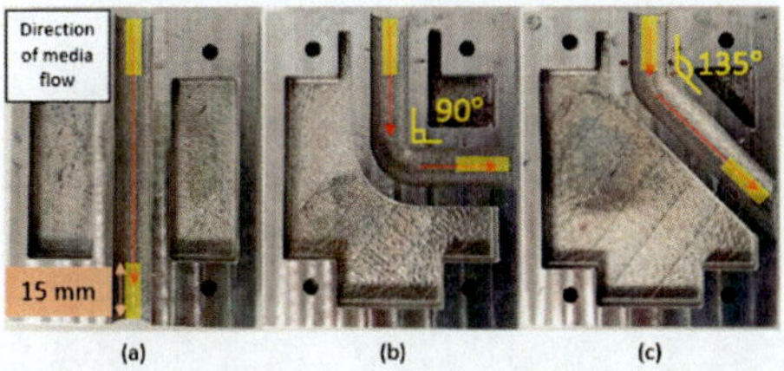

Fig. 1. (a) straight channel, (b) right-angled channel, (c) obtuse-angled channel. The yellow boxes highlight the areas used for line roughness measurements

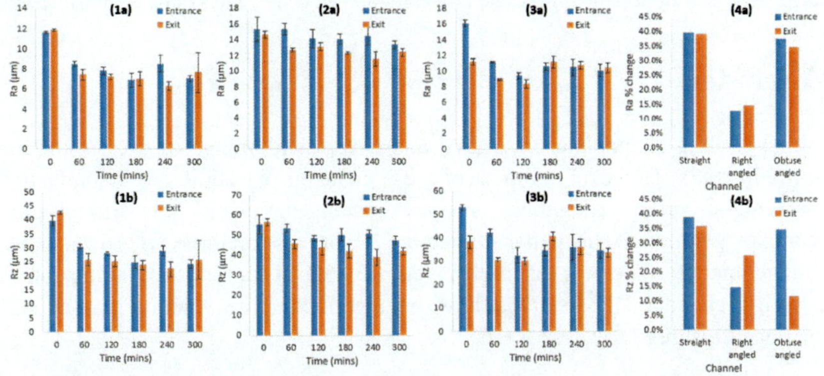

Fig. 2. Surface roughness characteristics (1a) Ra and (1b) Rz for the straight channel, (2a) Ra and (2b) Rz for the right-angled channel, (3a) Ra and (3b) Rz for the obtuse-angled channel, and summary of roughness changes to (4a) Ra and (4b) Rz

3.2 Simulation

Media Motion

The differences in finishing for the different channel geometries can be explained by the DEM simulation study. Simulations are conducted using Altair's EDEM software. The DEM model is simulated for 4 s. The media movement within the internal channels is shown in Fig. 3. Media motion within the channel can be classified into spiraling and back-forth motions. Spiraling is achieved when the media has high enough velocity to traverse through the channel. On the other hand, back-forth motion results when media does not have sufficient kinetic energy to traverse through the channel.

From the sliced views of the channels in Fig. 3, spiraling motion is observed only at the entrance of the straight and obtuse-angled channels. This explains why the Ra and Rz reductions in the straight and obtuse-angled channel are the highest. Due to the relatively slower velocity of media at the exit of the obtuse-angled channel, the exit of the obtuse-angled channel suffers from a poorer reduction in Rz than the straight channel. The right-angled channel is observed to have back-forth motion only, consequently observed to have the worst reduction in surface roughness out of the three channels. Since the back-forth media motion of the right-angled channel at the exit is faster than the entrance, there is slightly increased Rz reduction at the channel exit.

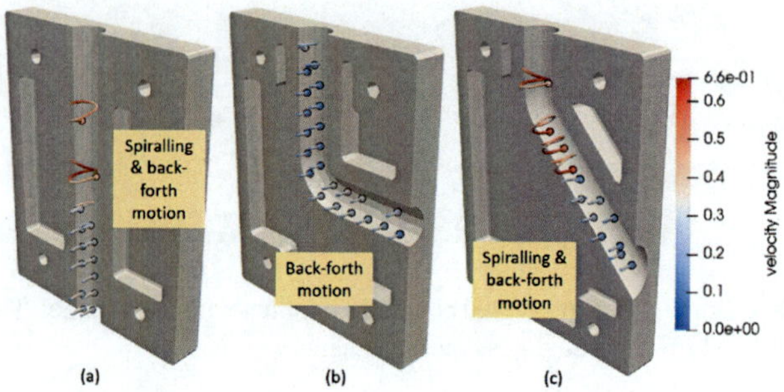

Fig. 3. Media flow in (a) straight channel, (b) right-angled channel, (c) obtuse-angled channel

4 Conclusion

Conformal cooling channels are essential and have a wide range of uses in the manufacturing industry. However, as most of these channels are additively manufactured, they have higher surface roughness. Surface finishing the channels is difficult due to their complex geometry. This paper studied the VF process on three SS316 channels of different geometries. The best result observed is 39.5% in Ra reduction at the entrance of the straight channel. DEM studies show that high media flow velocity is essential for good surface finishing.

References

1. Hamidi Nasab, M., Falzetti, A., Redaelli, A., Lecis, N., Giussani, A., Sala, L., et al.: Finishing of internal and external surfaces produced by powder bed fusion. SIG Addit. Manuf. **1**, 1–3 (2017)
2. Altaf, K., Rani, A.M.A., Raghavan, V.R.: Prototype production and experimental analysis for circular and profiled conformal cooling channels in aluminium filled epoxy injection mould tools. Rapid Prototyping J. **19**(4), 220–229 (2013)
3. Deepika, S.S., Patil, B.T., Shaikh, V.A.: Plastic injection molded door handle cooling time reduction investigation using conformal cooling channels. Mat. Today: Proc. **27**, 519–523 (2020)
4. Jahan, S.A., Wum, T., Zhang, Y., Zhang, J., Tovar, A., El-Mounayri, H.: Thermo-mechanical design optimization of conformal cooling channels using design of experiments approach. Procedia Manuf. **10**, 898–911 (2017)
5. Saifullah, A., Masood, S., (eds.).: Finite element thermal analysis of conformal cooling channels in injection moulding. In: Proceedings of the 5th Australasian congress on applied mechanics, Engineers Australia (2007)
6. Dimla, D., Camilotto, M., Miani, F.: Design and optimisation of conformal cooling channels in injection moulding tools. J. Mater. Process. Technol. **164**, 1294–1300 (2005)
7. TCT Magazine.: Post-processing of additive manufactured parts with cooling channels (2020). https://www.tctmagazine.com/additive-manufacturing-3d-printing-news/post-processing-additive-manufacturing-components/
8. Pakkanen, J., Calignano, F., Trevisan, F., Lorusso, M., Ambrosio, E.P., Manfredi, D., et al.: Study of internal channel surface roughnesses manufactured by selective laser melting in aluminum and titanium alloys. Metall. Mater. Trans. A. **47**(8), 3837–3844 (2016)

Rapid and Stable Plasma Transformation of Polyester Fabrics for Highly Efficient Oil-Water Separation

Ye Sun[1(✉)], Bo Ouyang[2], Rajdeep Singh Rawat[3], and Zhong Chen[1]

[1] School of Material Science and Engineering, Nanyang Technological University, Singapore 639798, Singapore
suny0038@e.ntu.edu.sg, ASZChen@ntu.edu.sg

[2] Nanjing University of Science and Technology, Nanjing 210094, China
ouyangboyi@njust.edu.cn

[3] Natural Science and Science Education, National Institute of Education, Nanyang Technological University, Singapore 637616, Singapore
Rajdeep.rawat@nie.edu.sg

Abstract. Fabrics with special wettability have drawn growing attention in recent years in the area of oil-water separation due to their low cost, good flexibility and ease of handling. However, an efficient and fast method to enable the required wetting state on fabrics still remains a challenge. In this work, a one-step, rapid and chemical-free hydrogen plasma treatment is reported to prepare a superhydrophobic and oleophilic polyester fabric. The as-prepared fabrics display static water contact angle of 153.2° with excellent oil-water separation capability. Mechanism of the surface transformation is discussed through chemical analyses, which indicate a significant removal of carboxyl group from the pristine hydrophilic surface. This developed method is envisaged to be used for on-demand large-scale production of materials for emergency oil cleanup through either separation or selective adsorption.

Keywords: Rapid · Single step · Chemical free · Superhydrophobic · Oil-water separation

1 Introduction

Great progress has been made in developing surperhydrophobic surfaces on various substrate materials including silicon [1], glass [2], and fabric [3]. Among them, fabric materials possess a vast potential for practical applications in separation of oil-water mixture or selective oil adsorption owing to their inexpensiveness and good mechanical flexibility [4]. Numerous methods of fabricating such superhydrophobic fabrics have been investigated. So far, most of the existing processes on modifying surface involve coating of low surface energy materials [5] such as modifying inorganic nanoparticles SiO_2 [6], TiO_2 [7] or using fluorocarbon silane [8] on a pre-roughened surface. These multi-step processes can be time-consuming [9] and may produce chemical waste. It is of

Y. Wei and S. Chng (Eds.): INCASE 2021, LNME, pp. 17–20, 2022.
https://doi.org/10.1007/978-981-16-5763-4_4

great interest to use a simpler physical process plasma technique to obtain the same effect in a single step with much reduced processing time. We can envisage that such capability will be useful for emergencies when materials have to be supplied quickly and on an on-demand basis. However, a simple one-step procedure to achieve the superhydrophobicity still remains a challenge. Here, we report a rapid, chemical-free and one-step process to realize superhydrophobic polyester fabrics (SPFs) via hydrogen plasma treatment. Excellent water contact angle (WCA) of 153.2° has been achieved after 4 min treatment. Such SPFs exhibit excellent separation efficiency of oil-water mixture. The mechanism of transforming originally hydrophilic surface to the superhydrophobic state is elucidated. The strategy provides a novel pathway towards practical applications such as treatment of oil contamination.

2 Methodology

We have prepared the SPFs via a radio frequency (RF) plasma system [10]. In a typical procedure, pristine polyester fabric sample (1.5 cm × 1.5 cm) was washed with ethanol and deionized water repeatedly, before being dried at 65 °C. The sample was then placed inside the RF-plasma system made of a quartz tube between the two capacitively coupled electrodes. Afterwards, the tube was evacuated to 10–3 mbar. Hydrogen gas then was introduced into the system at a flow rate of 10 sccm. The system stabilized at pressure of 0.45 mbar. H_2 plasma was created using 250 W RF power from a 13.56 MHz Caeser136 RF generator connected to the ring electrode through auto-impedance matching unit. The plasma treatment of the polyester fabrics was carried out for 4 min.

3 Results and Discussion

As-prepared SPF has a static water contact angle (WCA) of 153.2° and water shedding angle (WSA) of $12 \pm 1.5°$. The excellent flexibility and appearance of the polyester fabrics were well preserved after the plasma treatment. The plasma-treated fabrics have a slight decrease in the mechanical strength and elastic modulus. Such decrease is attributed to the change of chemical bonding state on the surface by the plasma treatment. Nevertheless, the slightly compromised mechanical properties will not affect our intended applications. Surface superhydrophobicity is typically derived from combined rough surface and surface chemistry affinity change against water. However, as observed under FESEM, the surface morphology of individual fibers in the SPF has no significant change after the plasma treatment. Therefore, the only change on the fabric surface comes from the surface chemistry modification (Fig. 1).

FTIR spectra showed apparent differences in the vibration peaks between 2800 cm^{-1} and 3000 cm^{-1} which are related to the carboxyl O-H group stretching. The reduction of carboxylic acid O-H groups peaks is responsible for the increased water resistance of SPF surface after the treatment. Further analysis through XPS was conducted to understand the surface chemical composition changes. A considerable reduction of oxygen peak intensity was observed after the plasma treatment, indicating strong oxygen removal. Significant reduction of O bonding (C-O and C=O) was detected. This is consistent with the FTIR observation that there was an apparent reduction of carboxyl group after

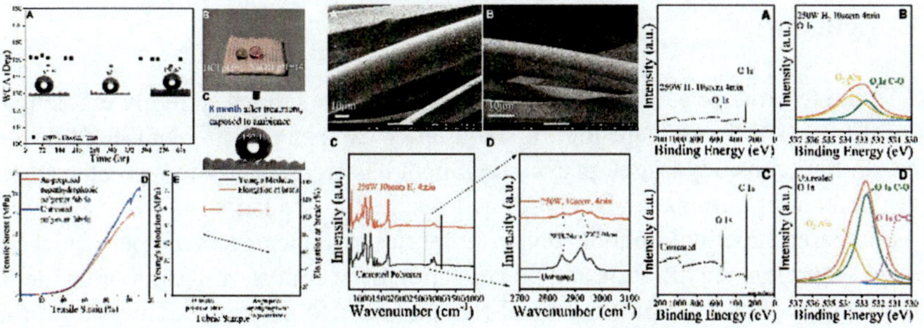

Fig. 1. WCA performance and mechanical stress comparison of SPFs; FESEM image and FTIR peak comparison; XPS spectra comparison.

the plasma treatment. Our results indicate that the mechanism for transformation of the superhydrophilic surface to superhydrophilic is due to the significant reduction of the surface carboxyl group after the plasma treatment. Correspondingly, the surface energy was substantially reduced.

Figure 2A and B show distinct difference when a piece of the plasma-treated fabric was placed on oil (Hexane, dyed with oil red for easy observation) and water (dyed with methylene blue). The treated fabric shows oleophilicity and hydrophobicity (floating on water, Fig. 2B).The treated fabric was used for gravity-driven separation of oil-water mixture. The fabric was placed at the nozzle of the funnel. Oil was able to pass through the porous fabric, but water could not (Fig. 2D). We calculated the oil separation efficiency for 10 cycles using e = (V – V0)/V, where V is the initial volume of oil and V0 the volume of oil lost during separation [11]. The superhydrophobic fabrics demonstrated a separation efficiency greater than 90% (up to 92%) for 10 cycles of filtration (the lost was due to the adsorption to the fabric). Furthermore, the as-prepared SPFs were examined for oil collection from oil-in-water surfactant-free emulsion (Fig. 2E). The emulsion was obtained by mixing water and hexane (dyed with oil red) in a volume ratio of 99:1, and then going through intensive stirring for 2 h [12]. The fabric was gradually dyed to red colour and the remaining mixture slowly became transparent. This suggests successful removal of oil from emulsion through adsorption.

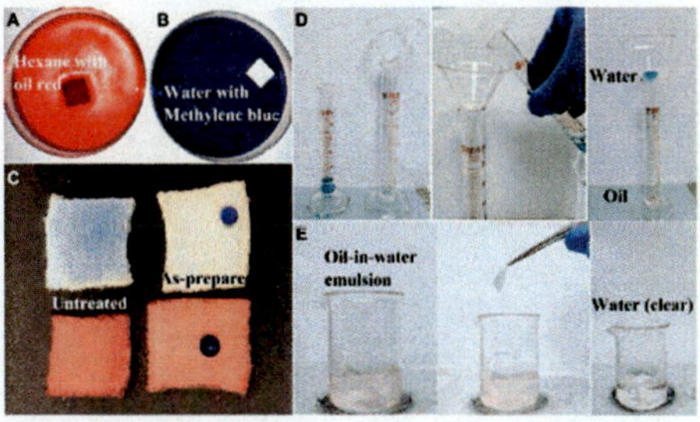

Fig. 2. Superhydrophobicity and oleophilicty of as prepared SPFs. Oil-water separation performance through filtration as well as adsorption.

4 Conclusion

Our Method to produce sueprhydrophobic and oleophilic fabrics is highly efficient and chemical-free. The fabrics are highly stable after exposure in ambient condition for more than 8 months. Hydrogen plasma treatment has significantly reduced the surface oxygen containing groups (C-O, C=O), and resulted in reduction of affinity to water. The as-prepared superhydrophobic polyester fabrics exhibited excellent performance in oil-water separation via filtration and oil collection from emulsion. The reported method has advantages of being rapid, chemical-free and easy to prepare, paving way for on-demand production of superhydrophobic fabrics for environmental cleanup and other applications.

References

1. Cao, L., Price, T.P., Weiss, M., Gao, D.: Super water- and oil-repellent surfaces on intrinsically hydrophilic and oleophilic porous silicon films. Langmuir **24**, 1640 (2008)
2. Mahadik, S.A., Kavale, M.S., Mukherjee, S.K., Rao, A.V.: Transparent superhydrophobic silica coatings on glass by sol–gel method. Appl. Surf. Sci. **257**, 333 (2010)
3. Ren, G., et al.: A simple way to an ultra-robust superhydrophobic fabric with mechanical stability, UV durability, and UV shielding property. J. Colloid Interface Sci. **522**, 57 (2018)
4. Xue, C.H., Bai, X., Jia, S.T.: Robust, self-healing superhydrophobic fabrics prepared by one-step coating of PDMS and octadecylamine. Sci. Rep. **6**, 27262 (2016)
5. Xue, C.H., Jia, S.T., Zhang, J., Tian, L.Q., Chen, H.Z., Wang, M.: Sci. Technol. Adv. Mater. **9**, 035008 (2008)
6. Zhou, C., et al.: Nature-inspired strategy toward superhydrophobic fabrics for versatile oil/water separation. ACS Appl. Mater. Interfaces **9**, 9184 (2017)
7. Hao, L., Gao, T., Xu, W., Wang, X., Yang, S., Liu, X.: Preparation of crosslinked polysiloxane/SiO2 nanocomposite via in-situ condensation and its surface modification on cotton fabrics. Appl. Surf. Sci. **371**, 281 (2016)
8. Zhang, W., Lu, X., Xin, Z., Zhou, C.: Nanoscale **7**, 19476 (2015)
9. Wang, H., Zhou, H., Gestos, A., Fang, J., Lin, T.: ACS Appl. Mater. Interfaces **5**, 10221 (2013)
10. Zhang, Y., et al.: Prereduction of metal oxides via carbon plasma treatment for efficient and stable electrocatalytic hydrogen evolution. Small **14**, 1800340 (2018)
11. Bano, S., Zulfiqar, U., Zaheer, U., Awais, M., Ahmad, I., Subhani, T.: Adv. Eng. Mater. 20 (2018)
12. Hui Liu, J.H., Chen, Z., Chen, G., Zhang, K., Al-Deyab, S.S., Lai, Y.: Chem. Eng. J. (2017)

Influence of Wettability of Micro-nano Structured Lead Bronze on Friction Properties

Lu Libin[1,3](✉) and Guan Yingchun[1,2,3]

[1] School of Mechanical Engineering and Automation, Beihang University, Beijing 100083, China
{lulibin,guanyingchun}@buaa.edu.cn
[2] National Engineering Laboratory of Additive Manufacturing for Large Metallic Components, Beihang University, Beijing 100083, China
[3] Advanced Manufacturing Center, Ningbo Institute of Technology, Beihang University, Ningbo 315100, China

Abstract. The micro-nano structure leads to changes in friction characteristics and surface wetting properties. Using a femtosecond laser, a micro-nano structure is constructed on the surface of the lead bronze. The laser-processed sample is soaked in absolute ethanol for 2 h to obtain a wettability that is different from the same morphology when it is not soaked. The water contact angle (WCA) measuring instrument is used to measure the WCA of different surfaces and the friction test is carried out with the UMT-3 rotating friction tester. Comparing samples with wetting properties under the same morphology, it is found that the change law of WCA and coefficient of friction (COF) is opposite. Increasing the surface WCA will improve the anti-friction performance. The larger the WCA, the better the sliding effect of the lubricating water film on the surface is, and the better it is to reduce the COF. This study is applicable to improve the antifriction performance of lead bronze.

Keywords: Femtosecond laser · Micro-nano structure · Wettability · Friction

1 Introduction

Micro-nano structures are usually used to improve the wetting and friction properties of materials. Wang et al. used a titanium-sapphire femtosecond laser to produce a textured surface with different microgroove spacing and tilt angles on the surface of AISI 304L steel. They found that the femtosecond laser-induced surface texturing significantly improved friction and wear performance [1]. Cao found the wide range of wettability is realized by controlled surface structure [2]. Generally, micro-nano structures lead to changes in friction characteristics and surface wettability [3–5], but there are few studies on the influence of the same structure's wettability on friction performance. Therefore, this paper uses a non-toxic and non-polluting femtosecond laser to construct micro-nano structures on the surface of the lead bronze, a commonly used bearing material, to explore the changing laws of its wettability and friction characteristics, and to study the effect of wetting performance on friction performance.

Y. Wei and S. Chng (Eds.): INCASE 2021, LNME, pp. 21–25, 2022.
https://doi.org/10.1007/978-981-16-5763-4_5

2 Methods

The lead bronze used in the experiment with the length and width of 10 mm and the thickness of 5 mm was grinded by 500, 1000, 1500, 2000, 2500 grid sandpaper. The mass fractions of Cu, Zn, Pb, and O are 54.37%, 42.5%, 2.33%, and 0.8% in sequence. The grinding part was a carbon steel ball (Shandong Haiou, China) with a diameter of 4 mm, and the surface was grinded to a surface roughness of 0.02 μm by mechanical grinding. The sample and the upper test piece were ultrasonically cleaned with acetone and absolute ethanol and then dried with compressed air. The surface texture was processed with a femtosecond laser system (PicoYL-15-0.1, Anyang Laser Technology Co., Ltd.) with a wavelength of 1064 nm, a pulse width of 209 fs, a frequency of 100 kHz, a focal length of 210 mm, and a focal point diameter of approximately 35 μm. The experimental parameters were set to 15 W laser power and 100 mm/s scanning speed. After processing, the surface impurities were ultrasonically cleaned with acetone and absolute ethanol and then dried with compressed air. The UMT-3 rotating friction tester (American Tribology Center) was used to carry out the friction experiment at a room temperature of 24 °C, a humidity of 23%, and a rotation diameter of 5 mm. The lead bronze plate was fixed on the platform and the carbon steel ball was fixed in the upper fixture. The samples before and after the friction test were ultrasonically cleaned with absolute ethanol for 5 min. To ensure the correctness of the experiment rules, the experiment was repeated three times.

3 Results and Discussion

As for the sample that had not been immersed in absolute ethanol, after laser processing, the WCA increased, but the COF decreased, as shown in Fig. 1. The grooves processed by the laser not only store wear debris but also store lubricants, forming a water film on the friction surface. Worn debris is easier to fall into the groove to avoid secondary friction and reduce the COF. After laser processing, the hydrophobicity of the surface is enhanced, and the WCA is increased from 89.7° to 101.1°. The better the hydrophobicity, the better the sliding effect of the lubricating water film on the surface is, and the better it is to reduce the COF [6–8]. When the unprocessed and processed samples were soaked in absolute ethanol for 2 h, the WCA decreased and the COF increased. Due to the same topography as the previous group, the laser-machined groove reduces the coefficient of friction. However, after soaking in absolute ethanol, the hydrophobicity of the laser processed sample surface is weakened, the WCA is decreased from 87° to 69.1°, and the COF increased. Therefore, the influence of wettability is greater than the influence of morphology, increasing the coefficient of friction [9, 10].

Figure 2 shows the friction morphology of various surfaces. It could be seen that the width of the wear scar of the sample after soaking in absolute ethanol was larger than that of the unsoaked sample, which was consistent with the COF after soaking as discussed above.

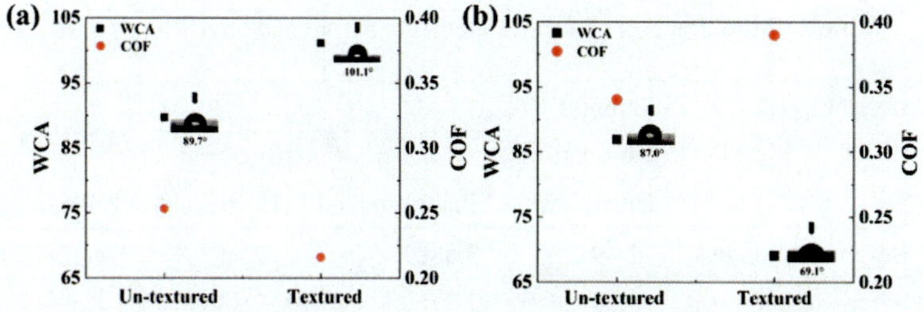

Fig. 1. Water contact angle (WCA) and coefficient of friction (COF) of different samples: (a) un-treated; (b) ethanol soaked.

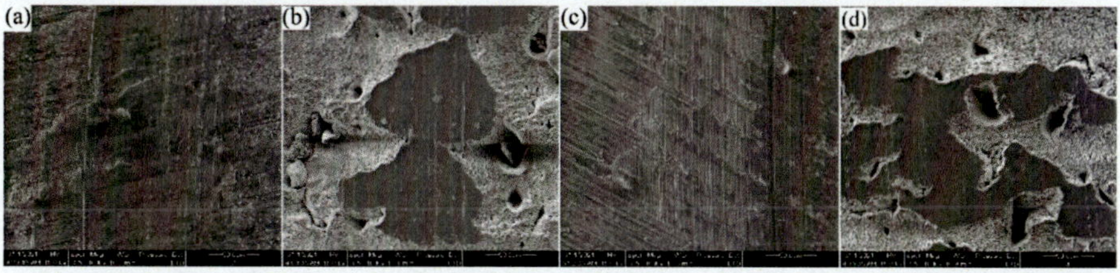

Fig. 2. Wear scar of samples (a) untextured, (b) textured, (c) untextured and ethanol soaked, (d) textured and ethanol soaked.

Table 1 shows the surface element mass fraction of the sample. During laser surface texturing, the thermal effect will lead to the Cu and Zn to be oxidized to form a dense oxide film, so the mass fraction of the O increased. Since the Zn and Pb are easier to react with O2 than the Cu, the mass fraction of the Zn and Pb increased. Moreover, the exposed Cu on the surface was reduced due to the dense oxide film covered on the surface, resulting in the mass fraction of the Cu on the surface reducing [11]. The produced micro-/nano-scale structures on samples are beneficial to fabricate superhydrophobic surface. Because that liquid can partially sit on air pockets trapped between the solid surface asperities leading to superhyrdrophobic condition [12]. It could be seen that the changing trend of the surface elements of the sample without ethanol soaking and after soaking was the same, but the lead content on the surface of the sample after soaking increased overall, which was more conducive to reducing COF. However, the actual COF was increasing, so the surface wettability played a greater role in the COF. This is mainly because that after soaking in absolute ethanol, the surface energy increases due to the change of surface element mass fraction, result in the decrease of WCA. It is known that, during friction test, a continuous layer of water can establish between pin and samples with hydrophobic surface, which have good effects on lubrication. Hence, the larger the contact angle, the smaller the COF [13].

Table 1. Mass fraction of surface elements before and after friction test

Mass fraction (%)		Samples			
		Un-textured		Textured	
		Un-treated	Ethanol soaked	Un-treated	Ethanol soaked
ω(O)	Before friction test	1.49	3.88	2.41	4.20
	After friction test	5.88	2.77	7.83	5.59
ω(Pb)	Before friction test	2.42	10.10	3.59	10.69
	After friction test	18.46	4.72	28.44	10.36
ω(Zn)	Before friction test	6.28	7.51	11.76	6.70
	After friction test	6.31	6.18	7.58	9.84
ω(Cu)	Before friction test	95.25	84.83	92.07	83.93
	After friction test	73.93	91.98	62.41	82.74

4 Conclusion

The change law of contact angle and COF was opposite. The larger the contact angle, the smaller the COF is, and the better the anti-friction performance was. This study confirmed the influence of surface wettability on friction performance and provided a reference for the research on anti-friction performance.

Acknowledgments. This work was supported by the National Key Research and Development Program of China (2018YFB1107700), National Natural Science Foundation of China (51875313).

References

1. Wang, Z., Li, Y.B., Bai, F., Wang, C.W., Zhao, Q.Z.: Angle-dependent lubricated tribological properties of stainless steel by femtosecond laser surface texturing. Opt. Laser Technol. **81**, 60–66 (2016)
2. Cao, L., Luo, B., Gao, H., Miao, M., Wang, T., Deng, Y.: Structure induced wide range wettability: controlled surface of micro-nano/nano structured copper films for enhanced interface. J. Mater. Sci. Technol. **84**, 147–158 (2021)
3. Wang, Y.Q., Luo, J.L., Heng, Y., Mo, D.C., Lyu, S.S.: Wettability modification to further enhance the pool boiling performance of the micro nano bi-porous copper surface structure. Int. J. Heat Mass Transf. **119**, 333–342 (2018)
4. Li, X., Zhang, L., Feng, Y., Zhang, X., Wang, D., Zhou, F.: Solid–liquid triboelectrification control and antistatic materials design based on interface wettability control. Adv. Func. Mater. **29**(35), 1903587 (2019)
5. Li, D., Wang, Z., Wu, D., Han, G., Guo, Z.: A hybrid bioinspired fiber trichome with special wettability for water collection, friction reduction and self-cleaning. Nanoscale **11**(24), 11774–11781 (2019)

6. Pan, Z., Peng, R., Tang, J., Chen, L., Cheng, F., Zhao, B.: Surface-segregation-induced nanopapillae on FDTS-blended PDMS film and implications in wettability, adhesion, and friction behaviors. ACS Appl. Mater. Interfaces. **10**(8), 7476–7486 (2018)
7. Zhao, X., Xiong, D., Liu, Y.: Improving surface wettability and lubrication of polyetherether-ketone (PEEK) by combining with polyvinyl alcohol (PVA) hydrogel. J. Mech. Behav. Biomed. Mater. **82**, 27–34 (2018)
8. Govind Rajan, A., Strano, M.S., Blankschtein, D.: Liquids with lower wettability can exhibit higher friction on hexagonal boron nitride: the intriguing role of solid–liquid electrostatic interactions. Nano Lett. **19**(3), 1539–1551 (2019)
9. Xing, Y., Wu, Z., Yang, J., Wang, X., Liu, L.: LIPSS combined with ALD MoS2 nano-coatings for enhancing surface friction and hydrophobic performances. Surfa. Coatings Technol. **385**, 125396 (2020)
10. Sun, Y., Guo, Z.: Recent advances of bioinspired functional materials with specific wettability: from nature and beyond nature. Nanoscale Horizons **4**(1), 52–76 (2019)
11. Huang, J., Guan, Y., Ramakrishna, S.: Tribological behavior of femtosecond laser-textured leaded brass. Tribol. Int. **162**, 107115 (2021)
12. Wang, H., He, M., Liu, H., Guan, Y.: One-step fabrication of robust superhydrophobic steel surfaces with mechanical durability, thermal stability, and anti-icing function. ACS Appl. Mater. Interfaces. **11**(28), 25586–25594 (2019)
13. Borruto, A., Crivellone, G., Marani, F.: Influence of surface wettability on friction and wear tests. Wear **222**(1), 57–65 (1998)

Check for updates

Effect of Ablative Layer Thickness on Laser Shock Peening Process Intensity

Thivyaa Ramesh(✉) and Niroj Maharjan

Advanced Remanufacturing and Technology Centre, Singapore 637143, Singapore
{thivyaa_ramesh,maharjan_niroj}@artc.a-star.edu.sg

Abstract. Sacrificial ablative layers are used in laser shock peening to protect the component surface from laser induced thermal damages. Factors such as thermal conductivity, absorptivity, acoustic impedance and thickness of the ablative layer play a critical role during the peening process. In this study, we investigated the effect of thickness of an ablative layer (black vinyl tape) on laser peening process intensity. Almen-C strips were peened with increasing thickness of ablative layer to observe changes in arc height deflection. It was found that increasing the thickness of the ablative layer from 130 μm to about 400 μm reduced Almen arc height deflection by more than half of its initial reading. This is attributed to the attenuation of shock wave intensity along the ablative layer due to longer propagation distance. The results suggest an inverse linear correlation between the thickness of ablative layer and laser peening process intensity.

Keywords: Laser shock peening · Ablative layer thickness · Almen strip

1 Introduction

Laser shock peening (LSP) is widely used to improve fatigue resistance of metal components in aircraft engines and nuclear power plants. During LSP, high energy laser pulses vaporize the surface to form a rapidly expanding plasma at the component surface, which is confined using a confinement overlay to generate high intensity shockwave through the component, introducing plastic deformation [1].

Generally, a sacrificial ablative layer is applied to the component surface during LSP to increase absorption of laser beam as well as prevent any thermal damage to the surface. Some of the commonly used ablative layers include black paint, aluminium/zinc tape and black vinyl tape. Contrasting results have been reported in literature on the effect of different ablative layer materials on peening [2–4]. This is generally attributed to the difference in ablative layer properties (such as absorptivity, impedance mismatch and conductivity). However, most studies use ablative layers of varying thicknesses for comparison which raises question regarding the explanation for contrasting results.

The thickness of the ablative layer could attenuate the compressive stresses being induced in the component [2]. However, no systematic study has been conducted to investigate this. Therefore, in this study, we specifically evaluate the influence of ablative layer thickness on peening by measuring Almen strip deflection. The deflection

Y. Wei and S. Chng (Eds.): INCASE 2021, LNME, pp. 26–29, 2022.
https://doi.org/10.1007/978-981-16-5763-4_6

corresponds to the magnitude of compressive stresses introduced by LSP. Results show that the thickness of ablative layer has inverse linear correlation with the Almen strip deflection.

2 Methodology

2.1 Almen Strip and Arc Height Deflection Measurement

LSP is a high intensity process, therefore Almen C strip (Grade C-1) was used to quantify the intensity of the process. Almen C strips were made of AISI 1070 cold rolled steel with dimensions as shown in Fig. 1. Pre-bow and post-bow Almen strip deflection measurements were made using TSP-3 #2 Almen gage. Both Almen gage and Almen strips are from Electronic Inc.

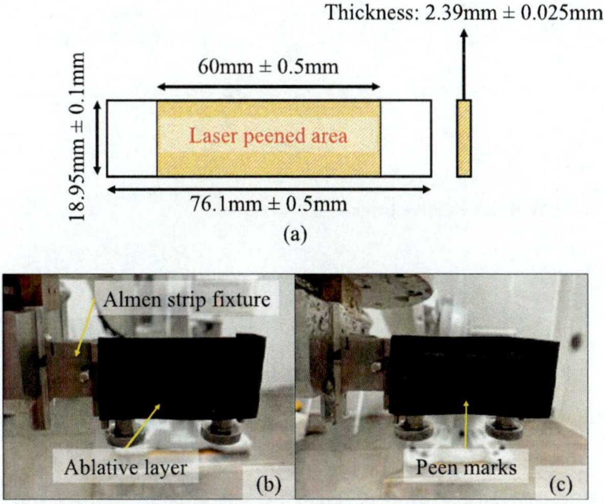

Fig. 1. (a) Almen C strip dimensions; (b) Almen C strip mounted on fixture before LSP; (c) Almen C strip after LSP.

2.2 Laser Shock Peening Trials

LSP was performed with a high-power Nd: YAG laser (Tyrida YS120-R200A), operating at 1064 nm wavelength. The laser delivers flat top pulses at 18 ns pulse duration with a frequency of 4 Hz. Laser output energy of 10 J and spot diameter of 3 mm was used (power density of 7.86 GW/cm^2). The Almen C strips were held on a fixture mounted on a 6-axis ABB robot arm. The laser peen pattern was designed to be registered in 4 steps to achieve an overlap of 50% and total coverage of 318%. 3M 471 black vinyl tape was used as an ablative layer. The tape thickness was varied from 130 μm to ~400 μm for the study. Three repeats of Almen C strip peening were performed with each thickness.

3 Results and Discussion

As shown in Fig. 2, an inverse linear relationship is observed between Almen C strip deflection and thickness of ablative layer used. Highest Almen C strip deflection of about 0.42 mmC is achieved when peened using ablative layer of low thickness (about 130 μm). This suggests a propagation of high intensity shock wave at Almen strip surface. Consistent with the inverse trend, lowest Almen strip deflection is observed for thicker ablative layer. There is also a good fitting of inverse linear correlation as indicated by large R^2 value. Almen strip deflection without any ablative layer is also shown in the graph for comparison. It can be seen that the absence of ablative layer significantly decreases Almen strip deflection.

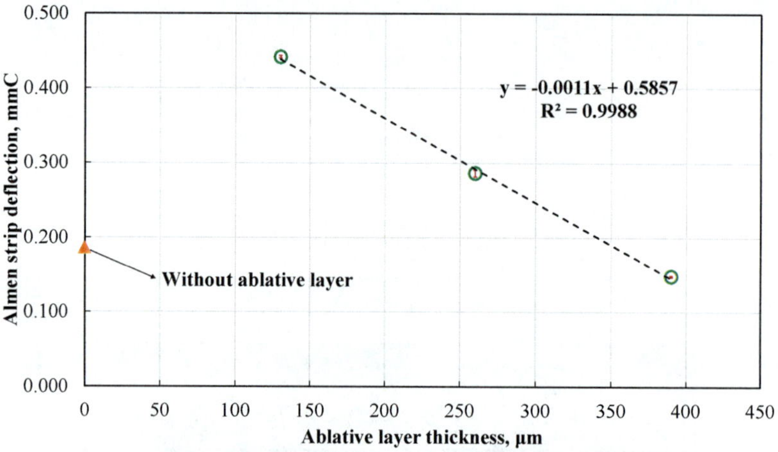

Fig. 2. Graph of Almen arc height values with varying ablative layer thickness.

Ablative layer aids in absorption of laser energy due to its opacity. In absence of ablative layer, less amount of laser energy is absorbed by the metal surface which produces a smaller plasma plume and thus, lower deflection of Almen strip. The presence of ablative layer enhances beam absorption [4]. This produces a larger plasma plume which is confined by water layer resulting in higher intensity recoil shock wave. Therefore, an increase in Almen strip deflection is observed when ablative layer is used.

This study exhibited that thickness of ablative layer has a direct correlation with Almen deflection. When thicker ablative layer is used, the Almen deflection linearly decreases. A potential mechanism for this is schematically illustrated in Fig. 3. For larger ablative layer thickness, plasma induced shockwave has to travel a longer distance before reaching the component surface. This results in the attenuation of generated shock wave along the path due to distance-related losses. By the time it reaches the metal surface, the intensity is much lower, which produces lower deflection. The results show that a thinner ablative layer is beneficial in increasing shock wave intensity. However, care must be taken when using very thin layer as it might result in thermal damage to the component due to heating effect of the laser.

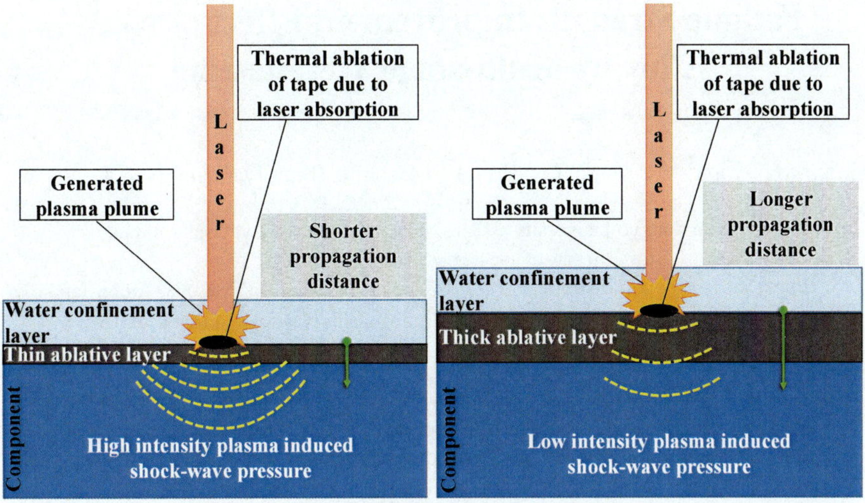

Fig. 3. Schematic of shockwave propagation with varying thickness of ablative layer.

4 Conclusions

A linear correlation was established between the ablative layer thickness and the Almen strip deflection. It was found that an ablative layer of 130 μm thickness produced highest level of Almen strip deflection. A thicker ablative layer results in pressure loss due to longer propagation distance for plasma-induced shock waves. The results show that thickness of ablative layer should be optimized to allow for effective shock wave pressure propagation without inducing thermal damages in component surface. Further studies on the effect of different ablative layer materials on peening are ongoing.

References

1. Maharjan, N., Chan, S.Y., Ramesh, T., Nai, P.G., Ardi, D.T.: Fatigue performance of laser shock peened Ti6Al4V and Al6061-T6 alloys. Fatigue Fract. Eng. Mater. Struct. (2021). https://doi.org/10.1111/ffe.13390
2. Peyre, P., Berthe, L., Scherpereel, X., Fabbro, R.: Laser-shock processing of aluminium-coated 55C1 steel in water-confinement regime, characterization and application to high-cycle fatigue behaviour. J. Mater. Sci. **33**(6), 1421–1429 (1998). https://doi.org/10.1023/A:1004331205389
3. Telang, A., Gill, A., Ramakrishnan, G., Vasudevan, V.: Effect of Different Ablative Overlays on Residual Stresses Introduced in IN718 SPF by Laser Shock Peening. Peening Sci. Technol. **1**, 75–86 (2018)
4. Yella, P., et al.: Laser shock peening studies on SS316LN plate with various sacrificial layers. Appl. Surf. Sci. **435**(2018), 271–280 (2018). https://doi.org/10.1016/j.apsusc.2017.11.088

Fatigue Strength Improvement Effect of AZ31 Alloy by Multi-stage Wet Blasting

Koichiro Nambu[1(✉)], Kenta Takashima[2], Masahiro Okumiya[2], and Isao Kumagai[3]

[1] Osaka Sangyo University, 3-1-1, Nakagatio, Daito-city, Osaka, Japan
knambu.mech@ge.osaka-sandai.ac.jp
[2] Toyota Technological Institute, 2-12-1, Hisakata, Tenpaku-ku, Nagoya, Aichi, Japan
[3] Macoho Co., Ltd., 525 Kanawa, Isurugi-machi, Nagaoka, Niigata, Japan

Abstract. Magnesium alloys have a problem of low corrosion resistance and fatigue strength. In this study, we focused on multi-stage wet blasting (WB) as a method to improve the fatigue strength. This method is characterized by a smaller increase in surface roughness compared to ordinary WB. We compared normal WB with multistage wet blasting. Magnesium alloy AZ31 was used as the test specimen. Fatigue tests were performed by rotating bending fatigue test. In the multistage wet blasting, particles with large particle size were treated, and then particles with small particle size were used. As a result, the fatigue limit of WB125 and WB15, which are multi-stage treatments, increased by about 10% compared to the normal treatment. It was found that this was due to a 16% smaller surface roughness as well as a deeper surface modified layer. Furthermore, it was reconfirmed that the particle size used in the first stage was the dominant factor in imparting hardness and residual stress in the wet blasting process.

Keywords: AZ31 · Fatigue strength · Multi-stage wet blasting

1 Introduction

Magnesium is lightweight and has high specific strength, so it is used in aviation, automobile parts, PC housings, etc. to reduce weight. However, it is used in limited places due to its low corrosion resistance and fatigue strength. In addition to this, it is difficult to use surface modification methods such as heat treatment and shot peening, which are generally used as methods for improving the fatigue strength of steel materials. In response to this problem, this research group is focusing on surface modification treatment using wet blasting treatment using compressed air for a mixed liquid slurry of water and abrasives. Previous studies have shown that wet blasting has a low fatigue strength improving effect due to the effect of surface roughness. As a new method, we focused on multi-stage wet blasting, which treats particles with different particle sizes in stages in order to prevent an increase in surface roughness and obtain a high fatigue strength improving effect.

The purpose of this study was to clarify the effect of multi-stage wet blasting on the fatigue strength characteristics of magnesium alloy AZ31.

Y. Wei and S. Chng (Eds.): INCASE 2021, LNME, pp. 30–34, 2022.
https://doi.org/10.1007/978-981-16-5763-4_7

2 Experimental Method

2.1 Materials and Specimens

Magnesium alloy AZ31 was used for material. The chemical composition of the alloy is 2.98% Al, 0.93% Zn, 0.3768% Mn, 0.0125% Si, 0.0013% Fe, 0.0013% Cu, 0.0005% Ni and Mg. The fatigue test specimens are shown in Fig. 1. The surface of the specimen was machined and then polished with emery paper #200 to 2000. The specimens were then annealed to relieve the internal stress generated during specimen preparation. The annealing was carried out by holding the specimens at 400 °C in air for 2 h and then cooling them in a furnace.

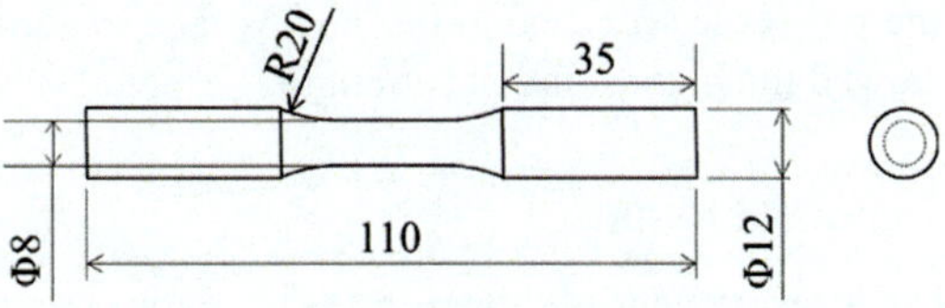

Fig. 1. Fatigue test specimen

2.2 Wet Blasting Conditions

The wet blasting equipment used was a baby blasting machine manufactured by Makoh Co. The projectiles used were FZB 120 (63–125 μm), FZB 205 (63 μm or less), and FZB 505 (10–30 μm). were used. The projectile pressure was 0.3 MPa and the projectile distance was 50 mm. The multi-stage processing conditions are shown in Table 1. The processing time for each stage was 60 s.

2.3 Various Experimental Methods

The fatigue test was a quadruple rotary bending fatigue test. The hardness was measured using a Micro Vickers hardness measuring machine. For the residual stress measurement,

Table 1. Multi-stage wet blasting process conditions

Name	process
AN	Annealed
WB 1	FZB 120
WB 2	FZB 205
WB 5	FZB 505
WB 125	FZB 120→FZB 205→FZB 505
WB 25	FZB 205→FZB 505
WB 15	FZB 120→FZB 505

a Cosα type residual stress measuring device manufactured by Pulstec Industrial Co., Ltd. was used. A stylus type surface roughness measuring device was used for surface roughness measurement.

3 Result and Discussion

3.1 Fatigue Strength Test Results

The results of the fatigue test are shown in Fig. 2. From this figure, it can be seen that the wet blasting treatment improves the fatigue strength. Furthermore, WB125 and WB15 showed the highest fatigue strength. Next, comparing the single-stage and multi-stage treated materials, the fatigue strength of the multi-stage treated material was higher. When compared to the projectile size used in the single-stage treatment, WB5 with the smallest grain size showed the highest fatigue strength.

3.2 Hardness Measurement Result

The result of hardness measurement is shown in Fig. 3. From this figure, we can see that the wet blast treatment increases the hardness. Next, comparing the hardness of the one-stage treatment and the multi-stage treatment, the multi-stage treatment has lower hardness than the one-stage treatment. Also, the depth of the hardened layer depends on the particle size used in the first stage. These results suggest that the particle size used in the first stage has a dominant influence on the hardness improvement. Furthermore, the reason for the lack of hardness improvement in the multistage treatment may be due to the large plastic strain in the first stage, which prevented the hardness improvement due to work hardening.

3.3 Residual Stress Measurement Result

The measurement results of residual stress are shown in Fig. 4. The dashed line in the figure shows the residual stress of the untreated material. The figure shows that the wet blast treatment imparts compressive residual stress. Next, a comparison of the one-stage and multi-stage treatments shows that, as in the hardness measurement results, the effect of the multi-stage treatment is small and does not differ significantly from the residual stress distribution of the one-stage treatment. This suggests that the particle size used in the first stage is also a dominant factor in adding compressive residual stress.

3.4 Surface Roughness Measurement Result

The measurement results of surface roughness are shown in Fig. 5. The surface roughness increases as the projectile grain size increases. Comparing the results of single-stage treatment and multi-stage treatment, it can be seen that the surface roughness decreases with multi-stage treatment. The decrease in surface roughness due to the multi-step process is more pronounced when particles with a large diameter are used in the first step. These results indicate that multi-stage wet blasting is an effective method for improving the fatigue strength of magnesium alloy AZ31, since it can reduce the surface roughness.

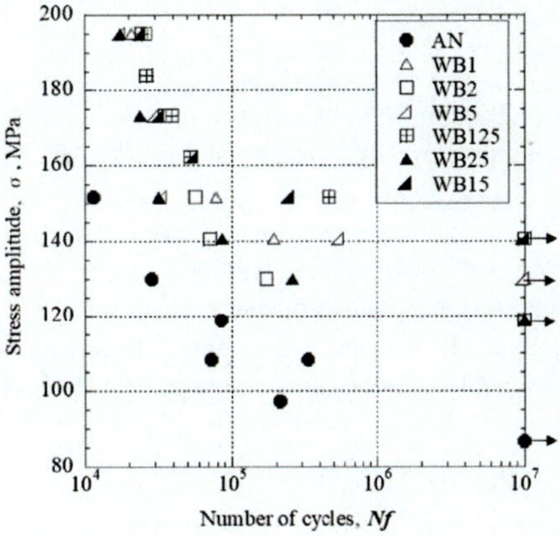

Fig. 2. Fatigue test result

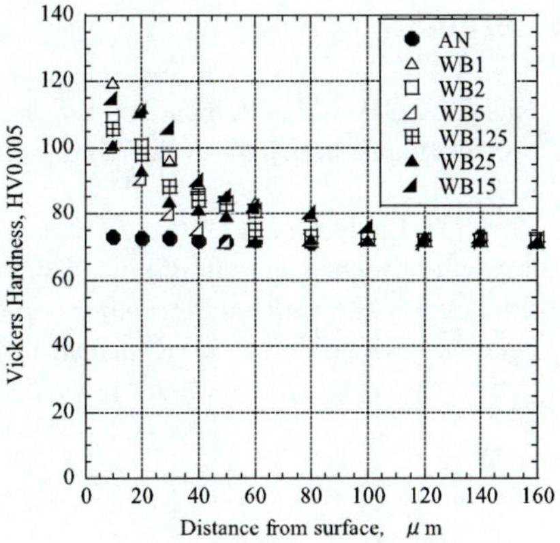

Fig. 3. Hardness distribution

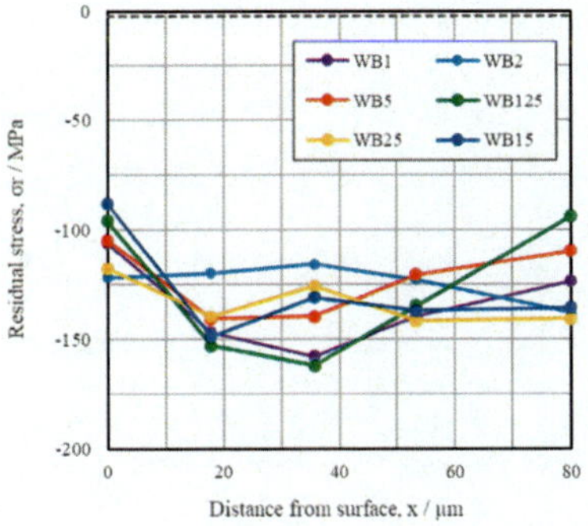

Fig. 4. Residual stress distribution

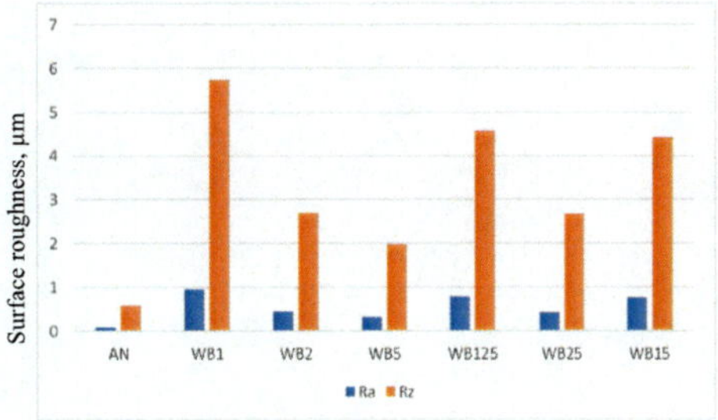

Fig. 5. Surface roughness

4 Conclusions

In wet blasting, the particle size of the particles used in the first stage is the dominant factor in imposing hardness and residual stress. Multi-stage blasting reduces surface roughness, but the change in hardness and residual stress is marginal. Multi-stage wet blasting is an effective way to improve the fatigue strength of magnesium alloy AZ31. Therefore, under the present conditions, WB125 or WB15 with some hardness and small surface roughness is effective for improving fatigue strength.

Acknowledgments. This work was supported by JSPS KAKENHI Grant Number 19K050840 and Contribute to the Development of LIGHT METALS RESERCH Grant.

References

1. Nakai, Y., Saka, M., Yoshida, H., Asayama, K., Kikuchi, S.: Fatigue crack initiation site and propagation paths in high-cycle fatigue of magnesium alloy AZ31. Int. J. Fatigue **123**, 248–254 (2019)
2. Zhang, P., Lindemann, J.: Influence of shot peening on high cycle fatigue properties of the high-strength wrought magnesium alloy AZ80, vol. 52, pp. 485–490 (2005). Author, F.: Contribution title. In: 9th International Proceedings on Proceedings, pp. 1–2. Publisher, Location (2010)
3. Nambu, K., Takashima, K., Kumagai, I., Okumiya, M.: Influence of Wetblast Treatment on Fatigue Strength of Magnesium Alloy AZ31. Advanced Surface Enhancement pp. 19–25 (2019)

Deep Cold Rolling (DCR) as Potential Surface Enhancement Process for Mold Inserts and Cores

Henry Kuo Feng Cheng(✉), Roger Chan, Jiakun Hu, and Wai Luen Chan

Advanced Remanufacturing and Technology Centre, 3 CleanTech Loop, CleanTech Two, #01/01, Singapore 637143, Singapore
{henry_cheng,roger_chan,jeffrey_hu, chan_wai_luen}@artc.a-star.edu.sg

Abstract. Stavax (AISI420) is a commonly used stainless steel in the mold manufacturing industry. Stavax inserts and cores are typically vacuum-hardened to improve their wear resistance. Improvement in the surface hardness is critical for the performance and durability of a mold. Besides, these mold inserts and cores consistently experience cyclic mechanical and thermal loadings in their operations. In this work, the potential of deep cold rolling (DCR) process has been explored as a mechanical surface enhancement for Stavax steel. The surface roughness of the DCR-treated coupons is found to be stabilised ~0.7 μm in the transverse direction and ~0.025 μm in the longitudinal direction, respectively. The surface hardness of the coupons increases between 2% and 6%, after DCR. The surface hardening of DCR-treated coupons can be due to the induced compressive residual stresses. A significant amount of process-induced compressive residual stresses is observed in both longitudinal and transverse directions. Therefore, it is found that DCR not only increases the surface hardness but also smoothens the surfaces due to the combined effect of the generation of compressive stresses and the introduction of cold work at the surfaces of the materials.

Keywords: Deep cold rolling · Surface enhancement · Vacuum-hardening · Stavax · Mold insert · Mold core

1 Introduction

Surface wear, fatigue and thermal fatigue are the most common failures in the molding industry. During operations, mold surfaces experience serious frictional forces due to the plastics flow, causing wear in the mold inserts and cores with time. This surface wear will be more significant when plastics contains hard and solid inorganic fillers. Furthermore, cyclic mechanical and thermal loadings would result in fatigue failure causing very costly rework and downtime [1]. Stavax (AISI420) is a premium stainless steel for small and medium inserts and cores. Stavax components are usually vacuum-hardened after machining [2]. Any improvement in the surface properties would leads to a better performance and a longer durability of the mold. Therefore, the surface treatment

Y. Wei and S. Chng (Eds.): INCASE 2021, LNME, pp. 35–39, 2022.
https://doi.org/10.1007/978-981-16-5763-4_8

on Stavax inserts and cores is a critical step to improve their wear performance as well as to prolong their durability.

Deep Cold Rolling (DCR) is similar to roller burnishing in which hydrostatically controlled rollers or balls are employed to induce surface and subsurface deformation. DCR produces smoother surfaces, deeper compressive stresses and a moderate cold work than shot peening, while restricting both the crack nucleation and growth [3–5]. In this work, the potential of deep cold rolling (DCR) has been explored as a surface enhancement process on Stavax steel for the mold manufacturing applications.

2 Methodology

The material used for the study was commercially available vacuum-hardened Stavax steel with the dimensions of 55 mm × 20 mm × 15 mm (L × W × H). The coupons were machined and polished prior to the DCR treatment. The process parameters and their respective sample codes are listed in Table 1.

Table 1: Sample codes and their respective process parameters

S/N	Sample code	Ball diameter	Pressure	Overlap	Track-width
1	04-200	4 mm	200 bar	50% of track-width	0.461 mm
2	04-400	4 mm	400 bar		0.620 mm
3	10-200	10 mm	200 bar		0.662 mm
4	10-400	10 mm	400 bar		0.910 mm

Surface roughness measurements were conducted using a Taylor Hobson's Form Talysurf. Surface hardness was measured in Vickers' hardness using Innovatest's Falcon 500. Surface roughness and hardness measurements were conducted at 5 different locations on each specimen and the average values are reported. Residual stresses were measured by Stresstech's Prism® system, an advanced hole drilling technique.

3 Results and Discussion

Surface roughness of pre and post-DCR of the specimens are shown in Fig. 1. The longitudinal roughness were measured along the rolling direction while the transverse roughness – perpendicular to the rolling direction. After DCR treatment, the transverse roughness increases. However, the longitudinal roughness of the coupons decreases significantly.

In DCR, the surface roughness reduces with increase in ball diameter and increase in number of tool passes whereas surface roughness increases with increase in rolling force and higher initial roughness of the surface [6]. However, the final surface roughness of the surface treated by DCR tends to stabilise below 1.0 μm [3]. In this study, the average initial roughness is considerably low due to the polishing process. Therefore, the increase in ball diameters could reflect only in the decrease of surface roughness in the longitudinal direction. In the transverse roughness, the higher values are found at the higher rolling pressure of 400 bar. However, the final roughness of DCR-treated coupons falls within the requirements of mold-manufacturing industry [7].

Surface hardness of pre and post-DCR of the specimens are presented in Fig. 2. Increase in surface hardness can be observed in all DCR-treated coupons. The increase in surface hardness by DCR can be explained by work hardening effects. However, in very hard material states (e.g. hardened steel) at the higher rolling force (pressure), near-surface work softening by DCR can also be observed [3].

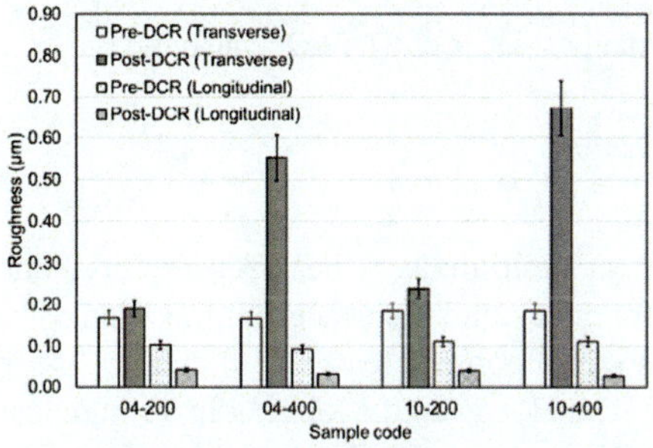

Fig. 1. Surface roughness of pre and post-DCR of Stavax coupons

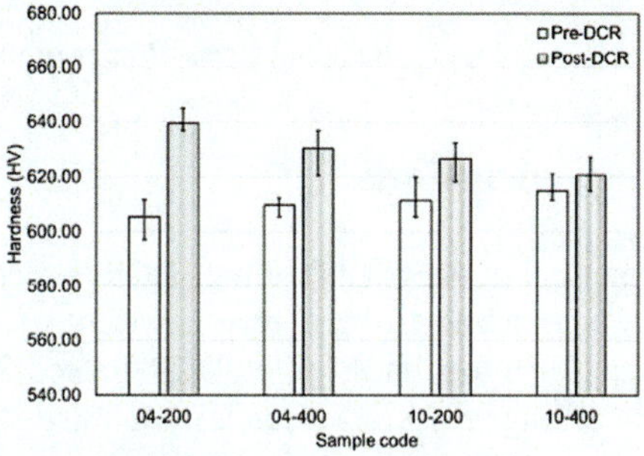

Fig. 2. Surface hardness of pre and post-DCR of Stavax coupons

In our case, the increase in the surface hardness is less significant at the higher rolling pressure of 400 bar compared to those rolled at 200 bar, reflecting on the near-surface work softening effect. Moreover, the DCR-induced work hardened near-surface layer would tend to particularly stablise in both mechanically and thermally to fatigue cyclic loadings owing to its finer grain size and higher dislocation density [5].

On the other hand, the surface hardening of DCR-treated coupons could also be due to an increase in compressive residual stresses [3]. The residual stress profiles of pre and post-DCR of the specimens are shown in Fig. 3. Significant amount of process-induced compressive residual stresses can be observed in both longitudinal and transverse directions. Typically, the compressive residual stresses in the longitudinal direction are smaller than that in the transverse direction, due to shear relaxation of residual stresses in the former direction.

Theoretically, DCR leads to maximum compressive residual stresses as governed by the Hertzian theory predicting maximum stresses below the surface [8]. From Fig. 3, increasing in the rolling pressure not only increases in the maximum compressive residual stresses at the sub-surface regions but also pushes the area of compression into greater depths. However, increasing in the rolling ball diameter only results in a deeper area of compression. This is because, with increasing rolling force (larger ball diameter with higher rolling pressure), the compressive residual stresses also increase until reaching to a saturated level. This saturated compressive residual stress is usually determined by the yield strength or work hardening state of the material. A further increase of rolling force shifts the area of compression into greater depths. However, too high rolling forces can lead to a reduction in the compressive residual stresses at the surface [3].

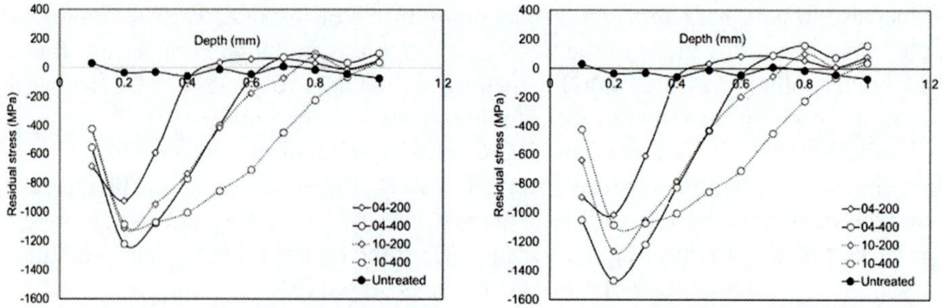

(a) Longitudinal residual stress **(b) Transverse residual stress**

Fig. 3. Residual stress profiles of pre and post-DCR of STAVAX coupons

4 Conclusions

Potential of deep cold rolling (DCR) process for mold industry has been explored and DCR is found to be a very powerful tool for mechanical surface optimisation. The transverse roughness of the DCR-treated Stavax coupons increases with an increase in the rolling pressure, yet tends to stabilise ~0.7 μm. However, the longitudinal roughness decreases significantly from ~0.1 μm to ~0.025 μm, after DCR. The surface hardness of the coupons increases between 2% and 6%, after DCR. The surface hardening of DCR-treated coupons could be due to an increase in compressive residual stresses. A significant amount of process-induced compressive residual stresses is observed in both longitudinal and transverse directions. Therefore, it can be concluded that DCR has a high potential as a surface enhancement process in the mold-manufacturing industry. However, dedicated assessments on both wear characteristic and cyclic mechanical and thermal fatigue should be studied.

References

1. He, B.: Research on the failure and material selection of Plastic mold. Procedia Eng. **23**, 46–52 (2011)
2. Stavax-ESR. https://www.assab.com/app/uploads/sites/133/2019/05/Stavax-ESR_EN.pdf
3. Altenberger, I.: Deep rolling—the past, the present and the future. In: Proceedings of 9th International Conference on Shot Peening, pp. 6–9 (2005)
4. Mader, S., Klocke, F.: Fundamentals of the deep rolling of compressor blades for turbo aircraft engines, modelling and simulation. In: Proceedings of 9th International Conference on Shot Peening, pp. 125–130 (2005)
5. Nalla, R.K., Altenberger, I., Noster, U., Liu, G.Y., Scholtes, B., Ritchie, R.O.: On the influence of mechanical surface treatments—deep rolling and laser shock peening—on the fatigue behavior of Ti–6Al–4V at ambient and elevated temperatures. Mater. Sci. Eng., A **355**(1–2), 216–230 (2003)
6. Prabhu, P.R., Kulkarni, S.M., Sharma, S.S.: An experimental investigation on the effect of deep cold rolling parameters on surface roughness and hardness of AISI 4140 steel. Int. J. Mech. Mechatron. Eng. **5**(12), 2639–2644 (2011)

7. Injection molding SPI surface finishes. https://www.hubs.com/knowledge-base/injection-molding-spi-surface-finishes/
8. Broszeit, E.: Improvement of fatigue properties by roll peening. Z. Werkstofftech **15**, 416–420 (1984)

Wettability and Spectrum Characteristics of Ti6Al4V Alloy Surface Based on Femtosecond Laser Texturing Technology

Bairui Du[1,2], Shengwang Tan[1,2], Bing Wang[1,2], Haiying Song[1,2(✉)], and Shibing Liu[1,2]

[1] Strong-Field and Ultrafast Photonics Lab, Faculty of Materials and Manufacturing, Beijing University of Technology, Beijing 100124, China
{hysong,sbliu}@bjut.edu.cn
[2] Key Laboratory of Trans-Scale Laser Manufacturing Technology, Ministry of Education, Beijing University of Technology, Beijing 100124, China

Abstract. In this paper, we used femtosecond laser texturing technology to prepare the LIPSS structure and the ablation pit structure on Ti6Al4V alloy surface by changing experimental processing parameters, we found that compared with the polished surface, when the laser fluence is lower than 0.1414 J/cm^2, the obtained LIPSS structure surface is easier to achieve hydrophobicity, and when the laser fluence is higher than 0.1768 J/cm^2, the obtained ablated structure surface is easier to achieve hydrophilicity. In addition, the spectral properties of the prepared surface micro-nano structures were tested. By comparing and analyzing the spectra, it was found that the reflectivity decreased with the increase of laser fluence in a certain range. When the laser fluence is 0.7070 J/cm^2, the light trapping structure is formed on the surface of Ti6Al4V alloy, so the reflectivity is reduced to about 2.5%. The results show that the selective regulation of spectrum can be realized by changing the characteristic scale of micro-nano structure on solid surface. It has great application potential in aerospace, biological materials, automotive energy and other fields. At the same time, it provides a certain technical basis for the development and application of surface modification of femtosecond ultrafast laser micro-nano technology materials.

Keywords: Femtosecond laser texturing · Ti6Al4V alloy · Surface wettability · Spectral selection characteristic

1 Introduction

In the field of material surface engineering, the surface micro/nano characteristics has brought many inspiration and enlightenment to human beings, such as the superhydrophobic structure on the lotus leaf surface [1], the light trapping structure on the butterfly wing surface [2], and the drag reduction structure on the shark skin surface [3]. This makes researchers realize that by changing the surface structure of materials, various surface properties of materials can be obtained to meet the needs of production

Y. Wei and S. Chng (Eds.): INCASE 2021, LNME, pp. 40–44, 2022.
https://doi.org/10.1007/978-981-16-5763-4_9

and life. Nowadays, after continuous exploration, a large number of surfaces with special properties have been widely used in self-cleaning, anti-corrosion, cell engineering, detection, solar cells and military stealth [4, 5]. The main methods for preparing micro-nano structural surfaces include template method [6], etching method [7] and electrochemical deposition method [8]. However, these methods have complex processing procedures, large processing costs, low efficiency, and large restrictions on processing size. In recent years, with the emergence of femtosecond laser texturing technology, a new tool is provided for the design and preparation of special material surfaces. Compared to the above methods, this technology has obvious advantages such as high processing accuracy, small processing size and wide processing materials. In this paper, direct ablation of Ti6Al4V alloy surface was induced by femtosecond laser texturing technology, and different micro-nano structures were prepared on the alloy surface. The variation of surface wettability and selective absorption characteristics of spectrum were monitored. In addition, the surface morphology was characterized and analyzed.

2 Methods

The size of Ti6Al4V alloy sample selected in the experiment was D25 mm × 10 mm, and the purity was 99.6%. The surface was polished and ultrasonically cleaned with anhydrous ethanol for 30 min. The cleaned samples were placed on a three-dimensional precision electronically controlled displacement platform (MFA-CC), the minimum moving distance of the platform is 0.125 μm. Set polyline "S" type path scanning, each polyline "S" is a period. The scanning speed and scanning interval are controlled by adjusting the moving speed and distance of the displacement platform, and the range of scanning speed is 0.1 mm/s–1 mm/s, the number of periods for each different processing parameter is 200. Then the micro-nano structures under different experimental processing parameters are obtained. The fs-laser system is a commercial product (Legend Elite, American Coherence Company) with 800 nm central wavelength, 60 fs pulse duration and a repetition rate of 1000 Hz. The laser source enters the OLYMPUS microscopic processing system, and focuses on the surface of the sample by the objective with numerical aperture N.A. = 0.25 and magnification of 10 times, and the spot diameter is 60 μm. The processing environment is air. SDC-350 contact angle analyzer was used to measure the surface contact angle (CA) between the sample and distilled water to evaluate the wetting behavior of the sample after patterning. The droplet was 2 μL, and the CA was the average of 5 measurements at different positions of the same sample. Carry 7000 UMS spectrophotometer (Agilent) was used in the experiment to study the reflectance spectrum of the sample surface.

3 Results and Discussion

In the experiment, the prepared samples were placed in air for 30 days and their CA was tested. The CA of the polished Ti6Al4V alloy was measured to be 46.68°. The scanning interval was set to 0.02 mm, and the scanning speed was set to 0.2 mm/s. The laser fluence was adjusted to 0.0707 J/cm^2, 0.1061 J/cm^2, 0.1414 J/cm^2, 0.1768 J/cm^2, 0.2828 J/cm^2, 0.3535 J/cm^2, 0.707 J/cm^2, and 1.0605 J/cm^2, respectively. The influence of the surface structure of Ti6Al4V alloy prepared at different laser fluences on CA was explored. The test results are shown in Fig. 1(a)–(h).

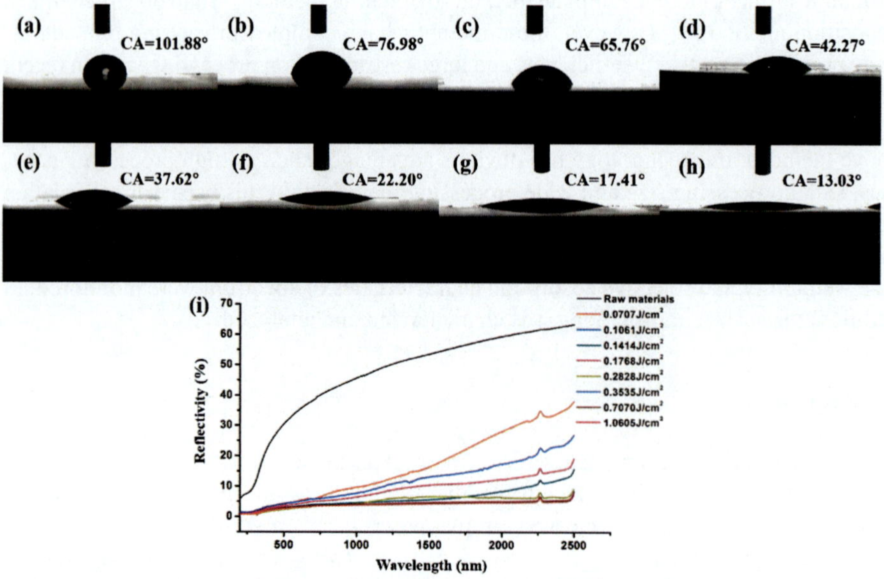

Fig. 1. CA on the surface of Ti6Al4V alloy obtained at different laser fluences. (a) 0.0707 J/cm^2; (b) 0.1061 J/cm^2; (c) 0.1414 J/cm^2; (d) 0.1768 J/cm^2; (e) 0.2828 J/cm^2; (f) 0.3535 J/cm^2; (g) 0.7070 J/cm^2; (h) 1.0605 J/cm^2. (i) Reflection spectra of Ti6Al4V alloy surface obtained at different laser fluences.

It can be seen from Fig. 1 that the surface structure constructed under different laser fluences has a great influence on the wettability of Ti6Al4V alloy surface. When the laser fluence is 0.0707 J/cm^2, the CA of Ti6Al4V alloy with micro-nano structure is 101.88°, and when the laser fluence is 1.0605 J/cm^2, the CA of Ti6Al4V alloy is 13.03°, showing hydrophilicity. Compared with the polished surface, when the laser fluence is lower than 0.1414 J/cm^2, the surface is easier to achieve hydrophobicity, and when the laser fluence is higher than 0.1768 J/cm^2, the surface is easier to achieve hydrophilicity. The spectrum of Ti6Al4V alloy surface with surface microstructure was measured. The results are shown in Fig. 1(i). The surface structure obtained under different laser fluences makes the reflection spectrum of Ti6Al4V alloy change significantly. In the 200 nm–2500 nm optical band, when the laser fluence is low, the absorption of light by the surface structure is also low. With the increase of laser fluence, the surface reflectivity decreases gradually. When the laser fluence is higher than 0.707 J/cm^2, the reflectivity decreases to about 2.5%, which means that the obtained surface structure can absorb most of the light. In addition, compared to the polished Ti6Al4V surface, it can be seen from the reflection spectra that all laser-processed surfaces exist a reflection peak near the wavelength of 2250 nm.

In order to further explore the reasons for the above phenomena, SEM was used to characterize and analyze the prepared surface structure. Figure 2(a)–(h) show the SEM

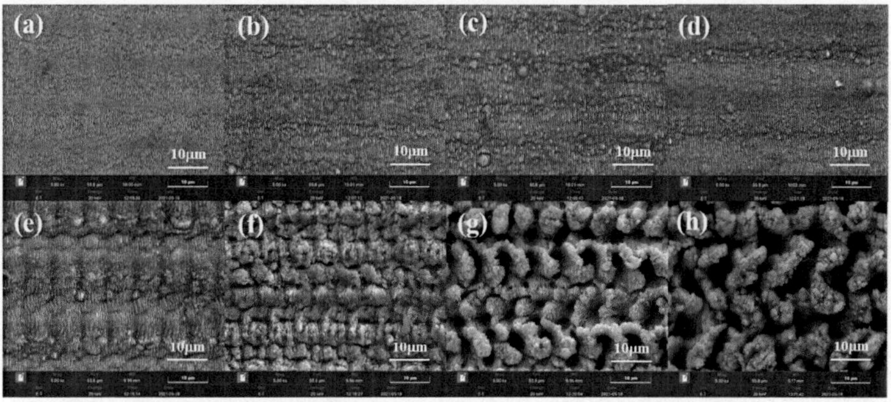

Fig. 2. SEM images of Ti6Al4V alloy surface obtained at different laser fluences.

images corresponding to the laser fluence of 0.0707 J/cm^2, 0.1061 J/cm^2, 0.1414 J/cm^2, 0.1768 J/cm^2, 0.2828 J/cm^2, 0.3535 J/cm^2, 0.707 J/cm^2 and 1.0605 J/cm^2, respectively. It can be seen from Fig. 2 that when the laser fluence is low, the laser-induced periodic surface structure (LIPSS) is generated on the surface of Ti6Al4V alloy, and the size of LIPSS increases with the increase of laser fluence. When the laser fluence is greater than 0.3535 J/cm^2, the ablation on the surface of alloy material is gradually obvious. When the laser fluence reaches 1.0605 J/cm^2, it can be clearly seen that an obvious multiple ablation pit structure is formed on the surface of the Ti6Al4V alloy.

For Ti6Al4V alloy, when the laser fluence is small, the LIPSS structure formed on its surface can make the alloy surface exhibit a certain degree of hydrophobicity. This is because the LIPSS structure forms a Cassie contact state between water and metal surface. When the laser fluence is relatively high, due to the formation of multiple ablation pit structure on the surface of Ti6Al4V alloy, the absorption capacity of the surface to water molecules is enhanced. So the surface wettability changes. In addition, due to the change of the surface microstructure morphology, it exhibits the characteristics of spectral selective absorption. When the laser fluence is low, the structural degree of Ti6Al4V alloy surface is low, so compared with the polished Ti6Al4V alloy surface, the reflectivity decreases with the increase of laser fluence. When the laser fluence reaches 0.7070 J/cm^2, the light trapping structure is formed on the surface of Ti6Al4V alloy, so the reflectivity is significantly reduced to about 2.5%. The periodic structure applied to the surface of the Ti6Al4V induces changes in the chemical and physical properties of the material surface under the action of the femtosecond laser, which further affects the change in the electronic energy level structure of the material, especially in the near-infrared wavelength range, this results in a reflection peak in the reflection spectrum near the wavelength of 2250 nm.

4 Conclusions

In this paper, the ablation of Ti6Al4V alloy surface was induced by femtosecond laser texturing technology. By adjusting the laser fluence, the LIPSS structure and the ablation pit structure were obtained respectively. Through the surface wettability test of these two structures, it is found that it is easier to achieve hydrophobicity on the surface of LIPSS structure with low laser fluence, and to achieve hydrophilicity on the surface of ablation structure with high laser fluence. This shows that the change of surface hydrophobicity can be controlled by adjusting the processing parameters. In addition, the spectra of these two surface structures were tested, and it was found that the reflectivity of the surface gradually decreased with the increase of laser fluence. When the laser fluence is high, the obtained ablation pit structure has light trapping characteristics, which can absorb most of the light in the 200 nm–2500 nm optical band, and combine with its excellent hydrophilic ability, which may have potential application value in many fields.

Acknowledgements. We gratefully acknowledge support for this work from NSAF of China (Grant Nos. 51705009 and 51875006).

References

1. Barthlott, W., Neinhuis, C.: Purity of the sacred lotus, or escape from contamination in biological surfaces. Planta **202**(1), 1–8 (1997)
2. Zheng, Y., Gao, X., Jiang, L.: Directional adhesion of superhydrophobic butterfly wings. Soft Matter **3**(2), 178–182 (2007)
3. Bixler, G.D., Bhushan, B.: Bioinspired rice leaf and butterfly wing surface structures combining shark skin and lotus effects. Soft Matter **8**(44), 11271–11284 (2012)
4. Dimitrakellis, P., Gogolides, E.: Hydrophobic and superhydrophobic surfaces fabricated using atmospheric pressure cold plasma technology. Adv. Coll. Interface. Sci. **254**, 1–21 (2018)
5. Yong, J., Yang, Q., Chen, F., Hou, X.: Femtosecond laser-induced superwetting surfaces (in Chinese). Chin. Sci. Bull. **4**(64), 1213–1237 (2019)
6. Celia, E., Darmanin, T., de Givenchy, E.T., Amigoni, S., Guittard, F.: Recent advances in designing superhydrophobic surfaces. J. Colloid Interface Sci. **402**, 1–18 (2013)
7. Liang, W., Zhu, L., Li, W., Liu, H.: Facile fabrication of a flower-like CuO/Cu(OH)2 nanorod film with tunable wetting transition and excellent stability. RSC Adv. **5**(48), 38100–38110 (2015)
8. Zhu, H., Gao, L., Yu, X., Liang, C., Zhang, Y.: Durability evaluation of superhydrophobic copper foams for long-term oil-water separation. Appl. Surf. Sci. **407**, 145–155 (2017)

Effect of Overlap in Robotic Hammer Peening on the Surface Integrity of Alloy Steel 4142

Wai Luen Chan[1](✉), Kai Hon Low[2], Henry Kuo Feng Cheng[1], and Roger Chan[1]

[1] Advanced Remanufacturing and Technology Centre, 3 Cleantech Loop #01-01, Singapore 637143, Singapore
{chan_wai_luen,henry_cheng,roger_chan}@artc.a-star.edu.sg
[2] NUS Mechanical Engineering, 9 Engineering Drive 1, Block EA #07-08, Singapore 117575, Singapore
low_kai_hon@u.nus.edu

Abstract. AISI 4142 is a commonly used alloy steel in machining tool due to its hardenability and low cost. In this study, Robotic Hammer Peening (RHP) is explored as a surface modification process to further enhance its material properties such as hardness and fatigue strength. The objective of this work is to examine the effect of overlap on the improvement of material properties of AISI 4142. Results revealed that increasing overlap produces increment in the surface hardness increases by up to 20%, and generates deeper residual stresses that can reach beyond 1.6 mm from the peened surface.

Keywords: Hammer peening · Overlap · 4142 steel · Surface integrity

1 Introduction

AISI 4142 is a low alloy carbon steel with medium hardenability. It is used for diverse applications in machining tools, such as bending dies, gears, spindles, and clutch parts. It is usually hardened to improve fatigue behaviour and desired strength using heat treatment [1]. Robotic Hammer Peening (RHP) is a mechanical surface modification process that can generate localized plastic deformation by means of high frequency oscillating movement [2]. RHP also exhibits the capability of creating a strain-hardening layer, which gives rise to stress-corrosion fatigue strength and wear resistance [3].

Several studies have been conducted to investigate the effects of different RHP process parameters. For instance, Mannes et al. have reported that the combination of lower impact force, smaller stroke length and deviation from 90° impact angle leads to reduction of compressive residual stresses of X3CrNiMo13-4 [4]. Hacini et al. also concluded that higher hardness and compressive residual stresses can be achieved with increasing impact energy level of hammer peening on 304L plates [5]. However, the influence of overlap on the material is still unclear. Therefore, fundamental investigations on the effect of different overlap during RHP on the surface integrity is discussed in this paper.

Y. Wei and S. Chng (Eds.): INCASE 2021, LNME, pp. 45–48, 2022.
https://doi.org/10.1007/978-981-16-5763-4_10

2 Experimental Setup and Methodology

AISI 4142 coupons with the dimensions of 60 mm × 60 mm × 20 mm were used, which were machined and polished prior to peening. RHP tests were conducted using motor-driven peening tool ECOpeen-A manufactured by ECOROLL AG, which was integrated to ABB industrial robot IRB 6660–205/1.9 model capable of 6-axis movement. The setup is shown in Fig. 1(a) and (b). The overlap, "*o*", is defined as the percentage of area overlaying between two neighbouring impacts, as shown in Fig. 1(c).

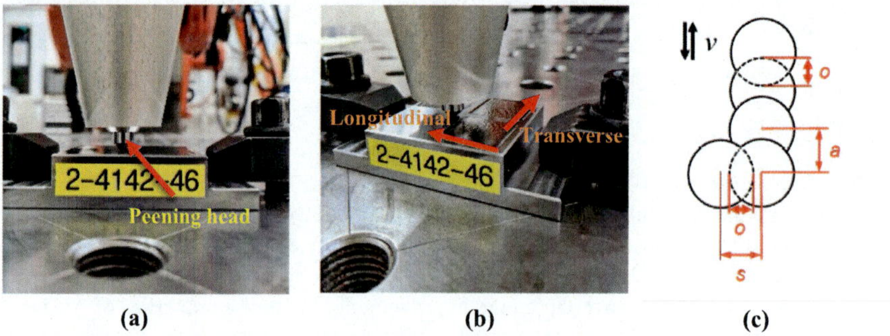

Fig. 1. RHP setup with (a) close view, (b) sample orientation, and (c) graphical illustration of overlap (*o*), indentation spacing (*a*) and pitch line distance (*s*).

Table 1. Experimental parameters

Parameters	Tool frequency	Impact energy	Stroke length	Tool radius	Overlap
Value(s)	220 Hz	80 mJ	2 mm	3 mm	400%, 1200%, 1600%, 2000%

The experimental parameters are listed in Table 1. The overlap settings were selected based on the current tool operational range. Innovatest's Falcon 510 was employed to evaluate the surface hardness (in Vickers) of the specimens with 0.5 kg load and 10 s dwell time. Residual stress measurements were performed using Stresstech's PRISM® system which is based on the principle of center hole drilling and speckle interferometry.

3 Results and Discussion

3.1 Surface Hardness

The surface hardness of as-received and post-RHP specimens with varying percentage overlap are illustrated in Fig. 2. As the overlap increases, the surface hardness of RHP treated specimens increases. The hardness increased up to 410 HV/0.5 with 2000% overlap. The hardness change increased from 7.51% to 24.37% as the overlap increased from 400% to 2000%.

The increase in surface hardness could be attributed to the increase in dislocation density after RHP process, which generates high plastic deformation [6]. Ultra-fine grains can be observed under the microstructure of high strength steel after RHP, which also promotes hardness improvement near surface layer [7]. As overlap increases, the contact between the RHP head and the material increases resulting in higher energy density from RHP process. This could lead to significant grain refinement in the microstructure. As a result, the strain-hardening layer becomes thicker, according to Hall-Petch strengthening relationship [8].

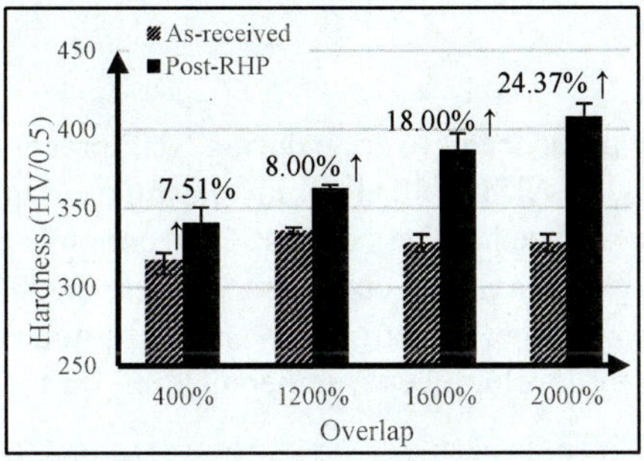

Fig. 2. Surface hardness of as-received and post-RHP specimens with different overlap at frequency 220 Hz and impact energy 80 mJ

3.2 Residual Stress

The residual stress profiles of as-received and post-RHP specimens in both longitudinal and transverse directions with varying percentage overlap are illustrated in Fig. 3. It can be observed that compressive residual stresses were imparted by RHP at all overlap settings in both directions, up to 1.6 mm depth. The measured residual stresses in transverse direction (normal to peening direction) are always higher compared to longitudinal direction (along peening path), due to the anisotropic behaviour of RHP process, as reported by Mannes et al. [6] and Bleicher et al. [9].

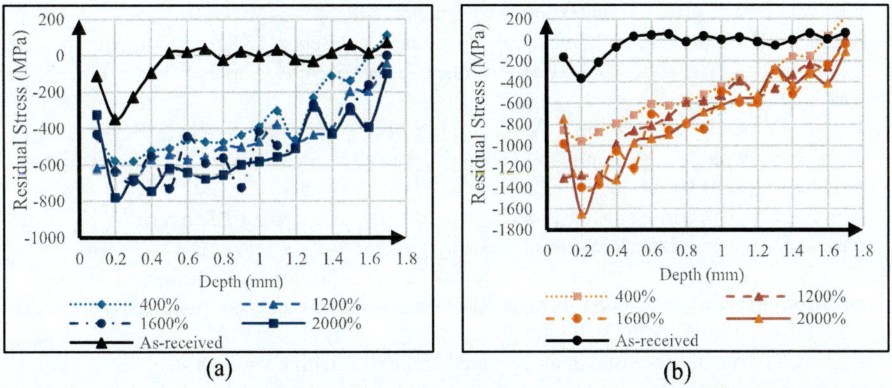

Fig. 3. Residual stress profiles of as-received and post-RHP specimens at different overlap in (a) longitudinal and (b) transverse directions

As the overlap during RHP increases, the maximum compressive residual stress increases in magnitude. The residual stress profiles also shift to the right with increase in depth of influence. This is because of the higher number of hammering impacts at large overlaps that give rise to higher energy density and Hertzian pressure generation [2]. Despite the impact energy from the current RHP tool set to be constant in all tests, the increase in contacts allow more kinetic energy to be transferred from the tool to the material to generate more plastic deformation. This eventually translates to enhancement of the fatigue life of treated components [10].

4 Conclusions

This paper examined the influence of overlap during RHP process on the residual stress state and surface hardness of AISI 4142 alloy. Increasing overlap resulted in more than 20% improvement in the surface hardness of post-RHP components. The residual stress profiles also exhibited higher maximum compressive stress magnitudes and deeper penetration (beyond 1.6 mm). Further investigations should be done on the microstructural analysis such as EBSD to develop deeper understanding on the impact of increasing overlap on grain structures.

References

1. Mckelvey, S., Fatemi, A.: Effect of Forging Surface on Fatigue Behavoir of Steels: A Literature Review. The University of Toledo, Ohio (2015)
2. Schulze, V., Bleicher, F., Grocche, P., Guo, Y., Pyun, Y.: Surface modification by machine hammer peening and burnishing. CIRP Ann. Manuf. Technol. **65**, 809–832 (2016)
3. Crocolo, D., Agostinis, M.D., Fini, S., Olmi, G., Paiardini, L., Robusto, F.: Tribological properties of connecting rod high strength screws improved by surface peening treatments. Metals **10**, 3434 (2020)
4. Mannens, R., Trauth, D., Mattfeld, P., Klocke, F.: Influence of impact force, impact angle, and stroke length in machine hammer peening on the surface integrity of the stainless steel X3CrNiMo13-4. Procedia CIRP **71**, 166–171 (2018)
5. Hacini, L., Le, N.V., Bocher, P.: Effect of impact energy on residual stress induced by hammer peening of 304L plates. J. Mater. Process. Technol. **208**, 542–548 (2008)
6. Mannens, R., Uhlmann, L., Lambers, F., Feuerhack, A., Bergs, T.: Surface integrity of AISI 52100 bearing steel after robot-based machine hammer peening. J. Manuf. Mater. Process. **4**, 61 (2020)
7. Revilla-Gomez, C., Buffiere, J.-Y., Verdu, C., Peyrac, C., Daflon, L., Lefebvre, F.: Assessment of the surface hardening effects from hammer peening on high strength steel. Procedia Eng. **66**, 150–160 (2013)
8. Yin, F., et al.: Ultrastrong nanocrystalline stainless steel and its Hall-Petch relationship in the nanoscale. Scripta Mater. **155**, 26–31 (2018)
9. Bleicher, F., Lechner, C., Habersohn, C., Kozeschnik, E., Adjassoho, B., Kaminski, H.: Mechanism of surface modification using machine hammer peening technology. CIRP Ann. Manuf. Technol. **61**, 375–378 (2012)
10. Kopsov, I.: The influence of hammer peening on fatigue in high-strength steel. Int. J. Fatigue **6**, 479–482 (1991)

Determination of the Interaction Depth Caused by Stress Peening at Leaf Springs

Eckehard Mueller(✉)

Bochum University of Applied Sciences, Am Hochschulcampus 1, 44801 Bochum, Germany
eckehard.mueller@hs-bochum.de

Abstract. Stress peening is a normal production step in leaf spring production. It enhanced the compressive residual stresses in the spring. The question that arises is what the amount is caused by shot peening and on the other hand coming from the prestress. A subtraction of the residual stresses along the leaf and perpendicular in dependence of the depth shows the increase. One result is a section that shows the limit of the maximal inducible residual stress. that is two-thirds of the tensile strength. The depth of the interaction is depended on the energy of the shots, the hardness of the shots, and the component to be shot peened. If the preload for peening is up to the tensile strength, depths of more than 500 μm are suitable for leaf springs. This method can be also done for coil or disk springs.

Keywords: Residual stress · Shot peening · Stress peening · Leaf spring

1 Introduction

Today the demand for light-weighted components in the automotive industry is more relevant than ever. Efforts are done for every single part. Springs are at the transition to sprung masses at a vehicle. It is stressed very high, and long durability is a basic demand. Therefore, many springs were not normal shot peened, but the next higher step stress peened to achieve high performance combined with lightweight.

2 Basics of Stress Peening Springs

Many components were shot peened to enhance the durability. It is a normal production process [1]. For some components, it is possible to stress peen them for additional enhancement of durability. In practice, it is done for five different kinds of parts. One part is a leaf spring that shows very easily the well-established process of stress peening described now [2].

The leaf spring has a preload during shot peening that gives a high amount of tensile stress. After unloading higher compressive residual stress is obtained in the surface layers. This procedure is called stress peening [3]. It works because the induced compressive residual stress by shot peening is independent of the stress before. This means, also for high tensile stress the same amount of compressive residual is received [4].

Y. Wei and S. Chng (Eds.): INCASE 2021, LNME, pp. 49–53, 2022.
https://doi.org/10.1007/978-981-16-5763-4_11

3 Experimental Setup

3.1 Leaf Spring Sample

Three parabolic leaf springs were examined. For the investigation, only one arm of the springs is used (Fig. 1). All specimens were produced on a normal production line with the normal decarburizing at the surface. Specimen 1 is the 2nd leaf of a two-leaf spring. Specimen 2 is the corresponding main leaf. Specimen 3 is a single-leaf spring. All relevant data of the springs are given in Table 1.

Fig. 1. The different specimens of parabolic leaf springs.

Table 1. Data of the springs.

Specimen	Material	Tensile strength R_m (MPa)
1	52 CrV 4	1800
2	52 CrV 4	1800
3	52 CrV 4 less P;S, Al	1665

Specimen 3 has less Phosphorus and Sulfur and also nearly no Aluminum. All specimens were coated after shot peening.

3.2 Residual Stress Determination

The residual stress determination was done by X-ray diffraction. A modern instrument from pulstec type μ-X360s was used. For determining the residual stresses, the cos-α-method was applied, see e.g. [5]. To get the residual stress profile the material was electrolytically removed. The measuring spot (4 mm diameter) was in the middle of the spring arm.

3.3 Peening Procedure

All specimens were stress peened. The relevant peening parameters were given in Table 2. It was done in the normal production line. The shot velocity was 80 m/s. The Almen intensity was greater than 0.4 mmA.

Table 2. Peening parameters.

Specimen	Preload during peening (MPa)	Shot size (mm)	Shot material
1	1800	1.2	Steel shot
2	1800	1.2	Steel shot
3	1870	1.0	Steel shot

4 Results

The compressive residual stresses were determined along and perpendicular to the leaf (Fig. 2 and Fig. 3). The values perpendicular show the compressive residual stress without preload. Along with the leaf, the residual is induced stress by stress peening.

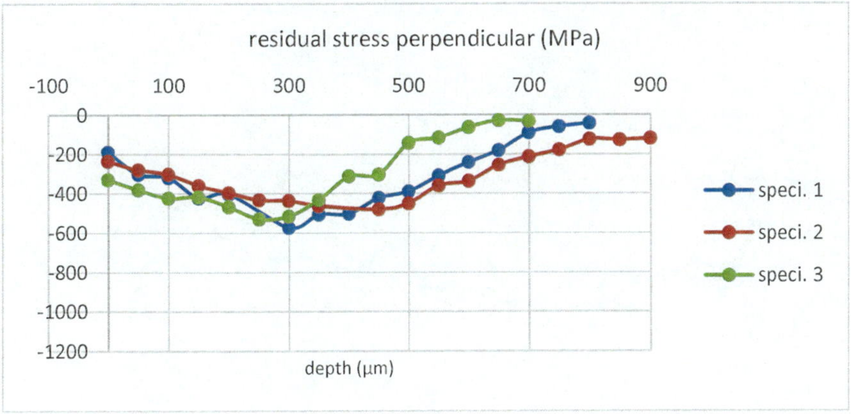

Fig. 2. Residual stress distribution perpendicular to the leaf

The difference between the residual stress along den perpendicular shows the amount of compressive residual stress additionally achieved by stress peening compared to normal peening (Fig. 4). The area from 0 µm to 200 µm is determined by decarburization of the material because row material without decarburization produces nobody. The flat part (dotted line) especially at specimen 2 shows that the limit of the induced residual stress is reached (here up to 700 µm). In deeper regions, several effects are mixed up. The leaf was scragged (plastified) before peening. Specimen 3 shows a smaller or nearly no flat behavior in the diagram caused by different shot intensities.

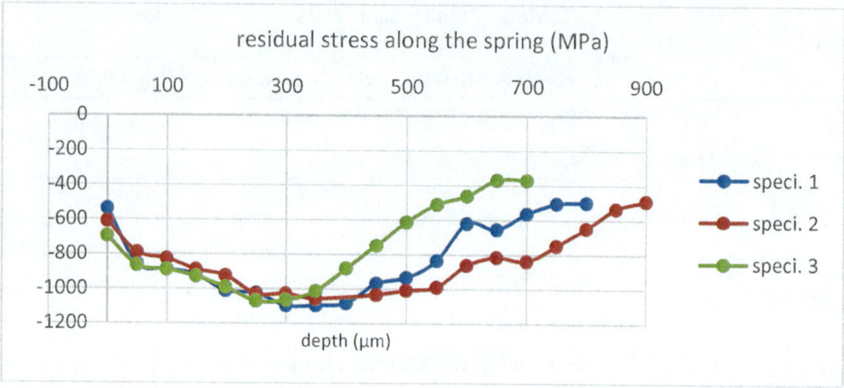

Fig. 3. Residual stress along the leaf

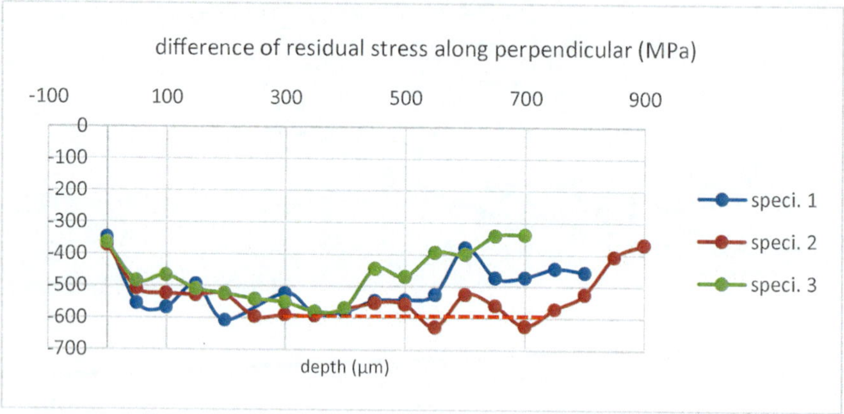

Fig. 4. Additional amount of compressive residual stresses by stress peening

5 Conclusion

Stress peening is a good method to enhance the compressive residual stress in a component by shot peening in two ways: The depth of penetration is much higher, and the amount can be increased up to the limit of the tensile strength R_m.

Acknowledgement. The authors would like to thank the research funding from the Research Fund Coal and Steel project GA-Nr 799787 called LIGHTTEC.

References

1. Baiker, S. (ed.): Shot Peening. 5th edn. MFN, Wetzikon, p. 2 (2017)
2. Mueller, E.: Stress peening as a performance enhancer of the normal shot peening process. MFN **5**, 26–28 (2004)

3. Baiker, S. (ed.): Shot Peening. 5th edn. MFN, Wetzikon, pp. 132–140 (2017)
4. Mueller, E.: The development of residual stress at bending specimens under the influence of setting and stress peening. In: Nakonieczny, A. (ed.) Proceedings ICSP, vol. 7, pp. 88–95. Institute for Precision Mechanics, Warsaw (1999)
5. Sasaki, T.: New generation X-ray stress measurement using Debye ring image data by two-dimensional detection. Mater. Sci. Forum **783–786**, 2103–2108 (2014)

Check for updates

New Electronic Peening Intensity Measurement Sensor Almen Intensity Correlation and Time Variation Study

Sylvain A. Forgues(✉), Brigitte Labelle, Adel Alouani, and Ramzi Ben Moussa

Shockform Aeronautique Inc., 4482 boul. de la Grande-Allee, Boisbriand, QC J7H 1R9, Canada
sylvain.forgues@shockform.com

Abstract. Shockform has heavily invested in R&D to develop a new electronic peening intensity measurement sensor. In this paper, the sensor is presented and its function is explained. Because companies calculate Almen intensity differently, the sensor output must be correlated to Almen intensity. An example of such a correlation is given for six different intensities. The experimental setup is illustrated and the peening parameters described. The correlation between sensor output and Almen intensity for one sensor is displayed. Using this correlation, the variation in sensor output is compared with the variation in Almen intensity for five different days over a one month period. The error for each day is calculated with respect to the initial value obtained on Day 1 and the average error for the five days is presented and discussed. Results show that the new electronic peening intensity measurement sensor is a good candidate to facilitate and accelerate the determination of peening intensity in a production environment.

Keywords: Electronic · Peening · Intensity · Measurement · Sensor

1 Introduction

Over the last decades, peening equipment has benefited greatly from the use of computers and robotic systems to improve the accuracy and the repeatability of the peening process while reducing the peening time. The same cannot be said of the method to measure peening intensity which has not changed significantly in decades. The current method requires the peening of several metal strips to different times in order to generate a saturation curve that represents a set of peening parameters. This method is very time intensive and the generation of a saturation curve yields different intensities depending on the supplier of the Almen strip, the number of points used in the saturation curve and the curve fit equation. The calculation to find the intensity using a mathematical method known as the 10% rule adds to the complexity since many operators find it difficult to understand.

A newly introduced wireless sensor is offering the possibility for near real-time peening intensity measurement, verification and transfer for storage and analysis. This sensor offers considerable time saving potential by eliminating all manipulation associated with Almen strips such as pre-bow measurement, installation on a holder, removal

Y. Wei and S. Chng (Eds.): INCASE 2021, LNME, pp. 54–57, 2022.
https://doi.org/10.1007/978-981-16-5763-4_12

after peening and arc height measurement. Furthermore, it is a green alternative since the sensor is reusable while Almen strips are discarded after a single use. This sensor is part of a new trend known as the Internet of Things (IoT) and promises to allow companies to progress into the latest industrial revolution called Industry 4.0 [1].

2 Sensor Description and Capabilities

The peening sensor has the size and screw pattern of a SAE J442 [2] Almen strip holder to easily be installed on most jigs and fixtures. The sensor was custom designed for this specific application.

Individual impacts are recorded during the peening process when media particles hit the load cell. Each impact has a different level of energy because of the size, hardness and velocity of the media particle. The impacts are recorded and sorted by a processor and a proprietary algorithm is used to calculate a sensor output voltage that correlates to Almen intensity. The sensor output is transferred via wireless communication to custom software on a Windows computer. Because the intensity is easy to measure, it becomes feasible to perform measurements during the peening of each individual component. This data can be recorded in a database for quality tracking and could eventually be used in a feedback loop to adjust peening parameters in real time.

Currently, sensors are available for A strip intensity range from 0.10 mmA (0.004 in. A) to 0.30 mmA (0.012 in. A). A lower intensity sensor for N scale peening is also available and a high intensity peening sensor is planned for the near future.

3 Correlation Between Sensor Output and Almen Intensity

Companies use Almen strips from different suppliers and generate saturation curves using different approaches and software which results in slightly different Almen intensities. Because of these differences, a correlation is required between the sensor output voltage and the Almen intensity. Correlating the sensor output voltage also ensures all sensors provide the same output.

To determine the correlation between sensor output and Almen intensity, saturation curves were generated using a SmartPeen® peening machine with full control and recording of the pressure and media flow rate. Five sensors and five Almen strips were

Fig. 1. Sensors and Almen strips in peening machine

placed alternatively on a turn table as per Fig. 1. The nozzle had a 9.52 mm (3/8″) internal diameter and was fixed at a distance of 6 in. The turn table was rotated at 4 rpm which translates to a tangential velocity of 4775 mm/min (188 in./min). Peening was performed at 6 pressures 0.41, 0.55, .83, 1.10, 1.38, 1.65 bar (6, 8, 12, 16, 20, 24 psi) at a media flow rate of 9.1 kg/min (20 lbs/min) using AWCR28 cut wire media meeting specification SAE AMS 2431/3D [3].

Almen intensities were determined from saturation curves generated by the Shockform Saturation Curve Solver (SSCS) [4] using the three parameter curve fit equation for five data points and the 10% rule as per SAE J443 [5]. The linear correlation between the Almen intensities and the output voltages for one sensor is illustrated in Fig. 2. The correlation is very good with a R^2 factor of 0.9947. The slope and the intercept of the correlation are entered in the personal computer software. With this information, the software will display sensor results in equivalent intensities as per the right column in Fig. 2.

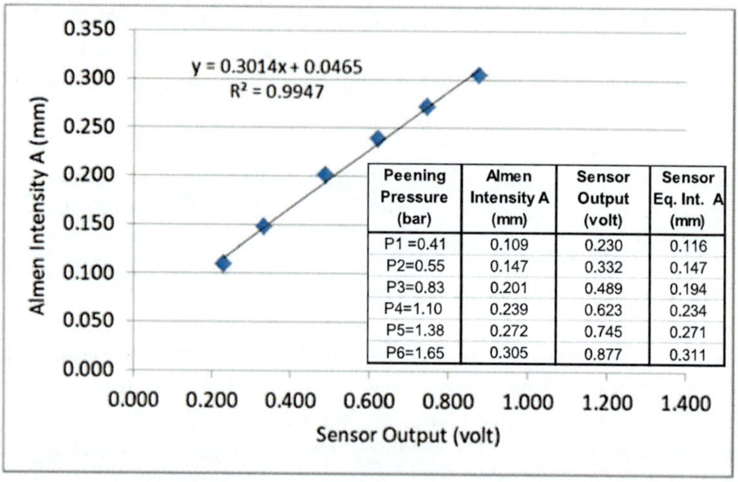

Peening Pressure (bar)	Almen Intensity A (mm)	Sensor Output (volt)	Sensor Eq. Int. A (mm)
P1 =0.41	0.109	0.230	0.116
P2=0.55	0.147	0.332	0.147
P3=0.83	0.201	0.489	0.194
P4=1.10	0.239	0.623	0.234
P5=1.38	0.272	0.745	0.271
P6=1.65	0.305	0.877	0.311

Fig. 2. Correlation between sensor output voltage and Almen intensity

4 Intensity Variation Over Time

The variation of Almen intensity and sensor equivalent intensity was studied over a period of 25 days. Almen intensity and sensor equivalent intensity was obtained from new saturation curves generated on day 5, 10, 15, 20 and 25. Intensity values from these new curves were compared with the baseline values shown in Fig. 2 and identified as Day 1. Figure 3 illustrates variation over time for both the Almen intensity (A-P1 to A-P6) and the sensor equivalent intensity (S-P1 to S-P6) where P1 to P6 are the pressures listed previously. The variation in sensor output over time is very similar to the variation in Almen Intensity.

To quantify this variation, the error in intensity for each data point was calculated with respect to the value obtained on Day 1. The average error for the five points was

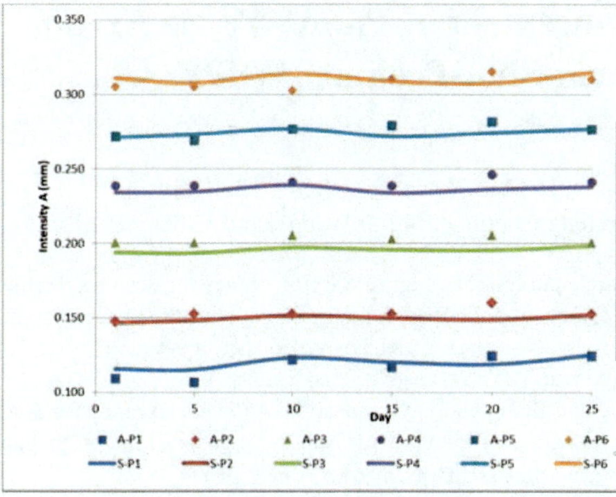

	Average Error Almen	Average Error Sensor
P1	9.8%	4.4%
P2	4.5%	2.3%
P3	1.3%	1.2%
P4	1.1%	1.0%
P5	2.2%	1.3%
P6	1.5%	1.0%

Fig. 3. Almen and sensor intensity variation with average error

then calculated and displayed in Fig. 3. The sensor error is similar or less than the error from Almen intensity. This shows that the sensor is as stable as the Almen strips to determine intensity.

5 Conclusion

A new electronic peening intensity measurement sensor was presented and its output was shown to correlate well with Almen intensity. Once correlated, the sensor provides a stable measure of peening intensity over several days with errors smaller or equivalent to the traditional method using Almen strips. This new sensor offers the possibility to save considerable manipulation time when determining or verifying intensity. It is also a much better choice for the environment.

References

1. Yang, et al.: Applications of Internet of Things in manufacturing. In: 20th IEEE International Conference (CSCWD 2016)
2. SAE J442: Test Strip, Holder, and Gage for Shot Peening. https://www.sae.org/standards/content/j442_202002/
3. SAE AMS 2431/3: Peening Media (AWCR), Conditioned Carbon Steel Cut Wire Shot, Regular Hardness. https://www.sae.org/standards/content/ams2431/3/
4. Shockform Saturation Curve Solver (SSCS), version 3.2. http://shockform.com/downloads32/
5. SAE J443: Procedures for Using Standard Shot Peening Test Strip. https://www.sae.org/standards/content/j443_198401/

Fatigue Enhancement of Ti-6Al-4V via *Ex-situ* Warm-Shot-Peening (WSP)

Zheng Zhang[1](✉), Debbie Hwee Leng Seng[1], Su Hui Lim[1,2], Augustine Kok Heng Cheong[1], Shijie Wang[1], and Jisheng Pan[1]

[1] Institute of Materials Research and Engineering, A*STAR (Agency for Science, Technology and Research), #08-03, 2 Fusionopolis Way, Innovis, Singapore 138634, Singapore
{zhangz,debbie-seng,augustine-cheong,sj-wang, js-pan}@imre.a-star.edu.sg

[2] Singapore Institute of Food and Biotechnology Innovation, A*STAR (Agency for Science, Technology and Research), #01-02, 31 Biopolis Way, Nanos, Singapore 138669, Singapore
lim_su_hui@sifbi.a-star.edu.sg

Abstract. To explore the potential beneficial effect beyond shot peening (SP)'s conventional performance, *ex-situ* warm shot peening (WSP) of Ti-6Al-4V coupons have been carried out. The coupons were pre-heated to 200 and 400 °C, before they were immediately transferred into a SP booth for shot-peening. While high cycle fatigue (HCF) tests did not detect a favorable improvement in fatigue life at high ($0.9F_{0.2}$) and medium loadings ($0.7F_{0.2}$), a significant increase in cycle-to-failure (N_f) in low loading ($0.5F_{0.2}$) was observed when the *ex-situ* WSP temperature was raised to 200 and further to 400 °C. Center hole drilling and electron backscatter diffraction (EBSD) were employed to unveil the mechanism for such fatigue improvement.

Keywords: Warm shot peening · Ti-6Al-4V · EBSD · Grain refinement

1 Introduction

Shot peening (SP) is a matured surface enhancement technique to extend the service lives of cyclically loaded metallic components, such as springs, suspension coils, etc. [1]. Through causing permanent plastic deformation at the surface layers, SP introduces compressive residual stress (CRS) at the top surface, which prevents surface micro-cracks from nucleation and propagation into the bulk of the components, therefore increasing the fatigue life of the components [2].

The beneficial CRS usually lays within a depth between 150 and 250 μm for most materials by using their respective optimized SP parameters. Recently, stress and heat have been coupled with SP in order to extend the beneficial effect of SP, with the aim of achieving a higher CRS value or a deeper layer with CRS. Some examples of surface modification include stress shot peening (SSP) [3], warm shot peening (WSP) [4], double shot peening (DSP), warm stress shot peening (WSSP), warm stress double shot peening

Y. Wei and S. Chng (Eds.): INCASE 2021, LNME, pp. 58–61, 2022.
https://doi.org/10.1007/978-981-16-5763-4_13

(WSDSP) [5, 6]. Among these modifications, WSP has an advantage of being applicable to components with irregular or asymmetrical geometries.

Inspired by the beneficial effect of WSP observed on SUP9 (Mn-Cr) spring steel [4], this work attempted *ex-situ* WSP on Ti-6Al-4V or Ti64 alloy, which is an important metal alloy widely used in automotive [7], aerospace [8] and medical [9] industries due to its high strength-to-weight ratio and excellent corrosion resistance. Ti64 coupons were heated up on a hotplate to 200 and 400 °C, before they were immediately transferred inside SP booth to carry out the traditional SP. The resultant fatigue life performance has been correlated with development in CRS distribution and microstructure in order to understand the mechanism behind the fatigue enhancement after *ex-situ* WSP at 400 °C.

2 Methods

Two geometries of commercial grade α/β Ti-6Al-4V (Ti64) coupons have been used in the current work (inset in Fig. 1(a)). The rectangular coupon had dimension of $50 \times 30 \times 10$ mm^3, while the flat dog-bone had a dimension of $124.3 \times 19.0 \times 2.5$ mm^3. The Ti64 coupons were heated on a Matcor MS7-H550-Pro precision hotplate. Their temperatures were measured using a thermal couple from the coupons' top surface. After the temperature was stabilized for 5 min, the coupon was immediately transferred onto a turning table inside the SP booth manufactured by Clemco Industries Corp. It took 3 min from transferring sample to starting of SP. High hardness cast steel shots with average diameter of 0.34 mm were peened at a pressure of 35 psi from a duckbill nozzle 4″ vertically above the Ti64 coupon surface. The nozzle was swiped 7 strokes across the coupon surface at a speed of 60 in./min in order to reach a target intensity between 6 and 9 A and a coverage of 125%. There was no vibropolishing after the *ex-situ* WSP.

The dog-bone coupons were used for high cycle fatigue (HCF) test on a MTS HCF Bionix 370 servohydraulic test system at three loads, equivalent to 0.9, 0.7 and 0.5 times of $\sigma_{0.2}$ at a frequency of 25 Hz and a load ratio (R) of 0.1 at room temperature. The rectangular coupons were used for the other analysis including average surface roughness (R_a) measurement via KLA Tencor P-10 surface profilometer and residual stress depth profile via center hole drilling (CHD) method. JEOL 7600F field emission scanning electron microscope (FE-SEM) equipped with X-Max 50 energy dispersive spectroscopy (EDS) detector and Oxford Instrument NordlysNano electron backscattering diffraction (EBSD) detector was used for microstructural characterization [10].

3 Results and Discussion

The HCF test shows that the cycle-to-failure (N_f) has significantly improved at all three loadings (Fig. 1(b)) after shot peening at room temperature without pre-heating of coupons (red circles). After heating the coupons up to 200 and 400 °C followed by SP, N_f at high and medium loads ($0.9F_{0.2}$ & $0.7F_{0.2}$) did not change significantly. However, the N_f at the low load of $0.5F_{0.2}$ increased clearly when the temperature was raised to 200 °C, and was further improved at 400 °C. This clearly differs from results in *in-situ* WSP study of Ti64, when SP and coupon heating were carried out concurrently [11].

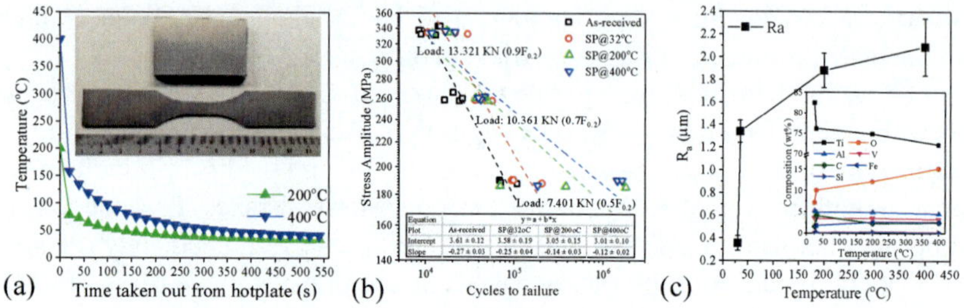

Fig. 1. (a) Temperature profiles of coupon surface after the coupons were removed from hotplate. The inset shows the photos of the rectangular and dog-bone coupons. (b) Stress-life (SN) curves from high cycle fatigue test and (c) coupon's surface roughness before and after shot peening at 32, 200 and 400 °C, with inset in (c) showing the surface composition (in wt%) by EDS.

The optimum *in-situ* WSP temperature was determined to be 100 °C which led to a maximum N_f at $0.5F_{0.2}$ loading above 10^6 cycles.

To understand the mechanism behind fatigue life enhancement after *ex-situ* WSP at 400 °C, a detailed analysis of the coupons was carried out. Surface roughness and O wt% of the coupons increased with *ex-situ* WSP temperature from 32 to 400 °C (Fig. 1(c)). The maximum compressive residual stress (CRS) was slightly smaller (−795.7 MPa) after *ex-situ* WSP at 200 and 400 °C as compared to that SP at 32 °C (−877.6 MPa). The layers with presence of CRS were of similar thickness (208 μm) for *ex-situ* WSP coupons and were much larger compared to that of as-received coupon (100 μm) (Fig. 2(a)). *Kernel average misorientation* (KAM) *maps* of the cross-section Ti64 surface before and after *ex-situ* WSP up to 200 °C showed only a very thin layer about 17 μm with greater misorientation (green layer) (Fig. 2(b)). However, after WSP at 400 °C, the green layer was much thicker at 32 μm. There was also appearance of unindexed (black) regions near the top edge, which were attributed to severe grain refinement [10, 11], which effectively delay the surface crack from nucleation and propagation inwards.

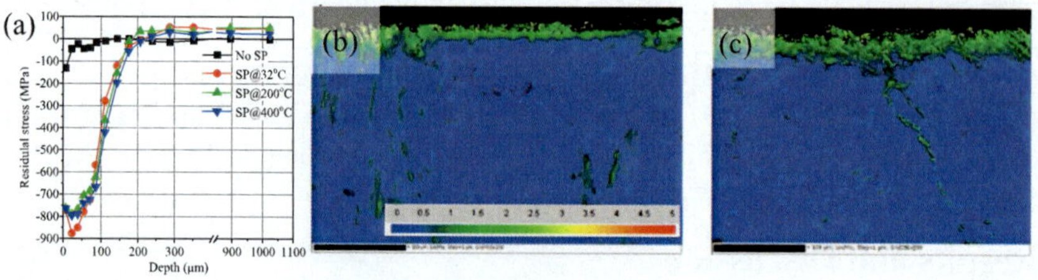

Fig. 2. (a) Residual stress depth profile by center hole drilling from Ti64 before and after shot peening at 32, 200 and 400 °C. Kernel average misorientation (KAM) maps from the cross-section Ti64 surface (b) before and (c) after ex-situ WSP at 400 °C.

Figure 1(a) shows that the Ti64 surface temperature dropped quickly after Ti64 coupons were taken away from the hotplate. The higher the heating temperature, the slower the cooling process and the higher temperature at the surface when the SP started. By heating up to 400 °C initially, the surface temperature cooled down to as high as

100 °C when the shot peening started at room temperature. From our earlier *in-situ* WSP study [11], SP at 100 °C brings the most beneficial effect. Hence, a good *ex-situ* WSP temperature was 400 °C to avoid excessive oxidation with improved grain refinement.

4 Conclusion

Ex-situ WSP of Ti64 coupons were carried out at 200 and 400 °C. An increase in *ex-siu* WSP temperature led to a higher surface roughness and higher surface oxidation. Meanwhile, higher initial heating temperature at 400 °C resulted in a higher surface temperature during shot peening, which induced much severe grain refinement at the surface observed under EBSD. Such enhanced grain refinement increased the fatigue life after *ex-situ* WSP at 400 °C, especially at lower load of $0.5F_{0.2}$.

Acknowledgements. This research is supported by the Agency for Science, Technology and Research (A*STAR) under its Aerospace Program (A1715a00079), Structural Metal Alloy Program (A18b1B0061) and Accelerated Materials Development for Manufacturing (A1898b0043).

References

1. Lainé, S.J., Knowles, K.M., Doorbar, P.J., Cutts, R.D., Rugg, D.: Microstructural characterisation of metallic shot peened and laser shock peened Ti–6Al–4V. Acta Mater. **123**, 350–361 (2017)
2. Jiang, X.P., Man, C.S., Shepard, M.J., Zhai, T.: Effects of shot-peening and re-shot-peening on four-point bend fatigue behavior of Ti-6Al-4V. Mater. Sci. Eng. A **468,** 137–143 (2007)
3. Wick, A., Schulze, V., Vohringer, O.: Effects of stress- and/or warm peening of AISI 4140 on fatigue life. Steel Res. **71**, 316–321 (2000)
4. Harada, Y., Mori, K.: Effect of processing temperature on warm shot peening of spring steel. J. Mater. Process. Technol. **162–163**, 498–503 (2005)
5. Scuracchio, B.G., de Lima, N.B., Schön, C.G.: Role of residual stresses induced by double peening on fatigue durability of automotive leaf springs. Mater. Des. **47**, 672–676 (2013)
6. Tange, A., Ando, K.: Improvement of spring fatigue strength by new warm stress double shot peening process. Mater. Sci. Technol. **18**, 642–648 (2002)
7. Faller, K., Froes, F.H.: The use of titanium in family automobiles: current trends. JOM **53**, 27–28 (2001)
8. Boyer, R.R.: An overview on the use of titanium in the aerospace industry. Mater. Sci. Eng. A **213**, 103–114 (1996)
9. Rack, H.J., Qazi, J.I.: Titanium alloys for biomedical applications. Mater. Sci. Eng. C **26**, 1269–1277 (2006)
10. Zhang, Z., et al.: Fatigue life enhancement in alpha/beta Ti-6Al-4V after shot peening: an EBSD and TEM crystallographic orientation mapping study of surface layer. Materialia, 100813 (2020)
11. Lim, S.H., et al.: *In-situ* warm shot peening on Ti-6Al-4V alloy: effects of temperature on fatigue life, residual stress, microstructure and mechanical properties. J. Alloys Compd. **882**, 160701 (2021)

The Research of Surface Roughness Prediction with Machine Learning According to Process Parameters in Laser Powder Bed Fusion

Jageon Koo, Eunju Park, Adrian Matias Chung Baek, and Namhun Kim(✉)

Department of Mechanical Engineering, Ulsan National Institute of Science and Technology, Ulsan 44776, South Korea
{wkrjsrn,stari12,adrianchung,nhkim}@unist.ac.kr

Abstract. Additively manufactured metal components are preferred in the industries that need to be reached high quality and durability. Although there are many advantages of freedom to design complex shapes and improved mechanical properties, low corrosion resistance and fatigue life still remain as challenges. One of the important factors of the challenges is surface roughness. A rough surface would wear and corrode more quickly than a smoother one. Especially, a down skin in an overhang structure has the highest surface roughness among the entire parts. To avoid rough surfaces, not only surface post processing but proper selection of process parameters is important. In this research, to decrease the down skin surface roughness of components fabricated by laser powder bed fusion (L-PBF), prediction of down skin surface roughness is conducted with machine learning algorithms in the step of selecting process parameters. The inputs of the data driven approach contain laser power (LP), scanning speed (SS), layer thickness (LT), hatching distance (HD) which are the main four factors in the L-PBF process, and overhang angles of specimens. Finally, through the data driven models, results show learning models are capable of predicting the down skin surface roughness of metal components fabricated.

Keywords: Laser powder bed fusion · Predictive model · Down skin surface roughness · Machine learning

1 Introduction

Metal additive manufacturing (M-AM), attributes to a freeform of fabricating even three-dimensional complex objects. Due to these characteristics, from simple tools to high quality industries, such that engines, aerospace, and energy components, metal AM have expanded its effectiveness [1]. Mechanical properties, one of the most important factors in determining high quality and lifetime, depends on withstanding high temperature and pressure or continuous repetitive loads [2]. Additively manufactured metal components resist higher tensile and compression loads than conventional manufacturing methods, but because of anisotropic microstructure and rough surface, fatigue life tends to be lower [3]. There are many studies to solve these hindrances, using thermal post processing and hybrid manufacturing techniques [4].

Y. Wei and S. Chng (Eds.): INCASE 2021, LNME, pp. 62–65, 2022.
https://doi.org/10.1007/978-981-16-5763-4_14

Especially, rough surface, that can easily causes wearing and corrosion, has been resolved by predicting surface quality with machine learning (ML) techniques for minimizing additional energy input [5–7]. Two main methods for prediction surface quality have been developed based on in-situ monitoring data and predicting based on process parameters. The method of surface roughness prediction through the process parameters has the advantage of being able to eliminate parameters that fail to manufacturing in advance, and this means contributing on the time and cost efficiency. In most cases, down skin of overhang structure or internal face, that shows the poor surface textures, should be considered more carefully for appropriate parameters [8].

This study, focused on the down skin surface of overhang structure, aims a prediction of the surface roughness from process parameters and overhang angles, using data driven algorithms in L-PBF. The remain parts will be followed by data acquisition, used ML algorithms result, and conclusion.

2 Methodology

In this research, CM247LC, which is superalloy derived from MAR-M247, in an overhang structure specimen was used for measuring down skin surface roughness. CM247LC powder was procured from Carpenter Additive Co., with particle sizes ranging from 15 μm to 45 μm. EOS M290 laser-powder bed fusion (L-PBF) machine was used in manufacturing process. In the process, main four parameters which are laser power (100–300 W), scanning speed (900–6000 mm/s), layer thickness (0.02 mm), and hatching distance (0.045–0.15 mm), with overhang angle (30–45°) were selected as input variables shown as Fig. 1. The input variables were selected in the range. In the range of input variable, a total number of 213 overhang surfaces of 71 samples were fabricated. The down skin surface roughness values of overhang structure samples were measured with Mitutoyo SJ-310 profilometer and the specimens are shown in Fig. 2. Surface roughness measurements are taken from 0.5 mm length at 5 different areas on given way in the figure. Figure 3 shows the measured down skin surface roughness values of overhang specimens. Use of these input data and output data employs machine learning regression models. Random forests regression (RFR), extreme gradient boosting (XGB), and support vector regression (SVR) were used learning algorithms. These algorithms have been widely used in manufacturing industries to predict time and quality.

3 Results

In this study, prediction of down skin surface roughness values of overhang samples was tried to forecast from process parameters with overhang angles with machine learning models. These features are fed into the regression algorithms as input variables. A total number of 213 overhang structure specimens were fabricated and measured for down skin surface roughness prediction. Random forest regression, extreme gradient boosting and support vector regression algorithms have been trained. The performances of each algorithms are given in Table 1. The results show that these proposed data-driven methods, using input variables are process parameters with overhang angle and output is down skin surface roughness, should be conducted more in detail for more application in other AM process.

4 Conclusion

RFR, XGB, and SVR will be more widely adapted in manufacturing industries through data accumulation and specific pre-processing. There are time and cost efficiency to do predicting quality and lifetime through surface roughness, before fabricating. But still a limitation exists in reducing down skin surface roughness, due to the technical problems of metal 3D printing that is surface post processing must be done. In the future, the proposed research is expected to high level of benefits for improved surface texture without surface post processing.

Table 1. Prediction performance of training algorithms

	MAE	RMSE	MAPE	R^2
RFR	1.054	1.378	0.601	0.550
XGB	0.947	1.242	0.550	0.634
SVR	1.300	1.586	0.748	0.406

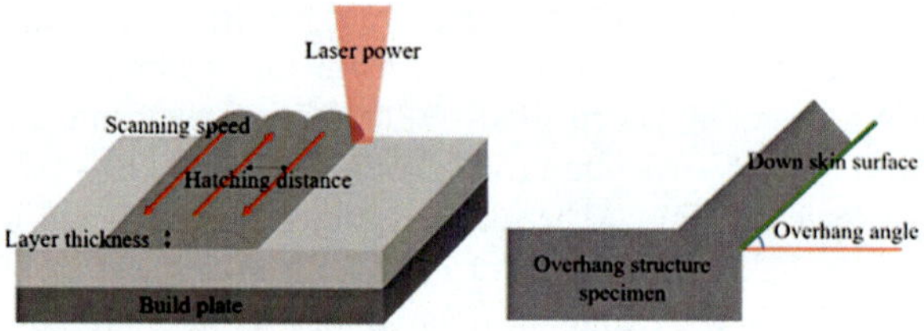

Fig. 1. Schematics for process parameters of L-PBF and means of overhang angle in overhang structure specimen

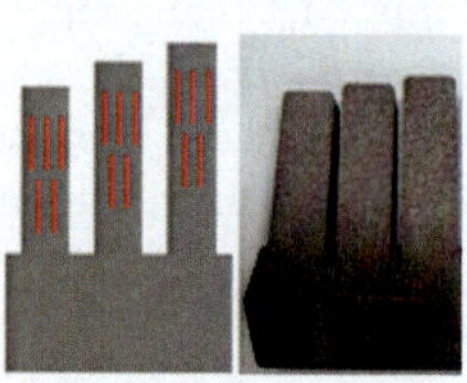

Fig. 2. Schematics of measurement and down skin of overhang specimen

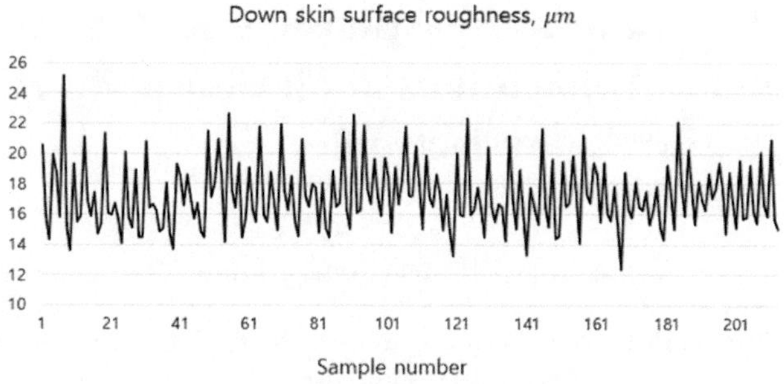

Fig. 3. Down skin surface roughness values of the samples

Acknowledgement. This work was supported by the Korea Institute of Energy Technology Evaluation and Planning (KETEP) grant funded by the Korea government (MOTIE). (20193310100030, Development of highly efficient F-class gas turbine hot component by controlling and applying Design for Additive Manufacturing).

References

1. Buchanan, C., Gardner, L.: Metal 3D printing in construction: a review of methods, research, applications, opportunities and challenges. Eng. Struct. **180**, 332–348 (2019)
2. Raffeis, I., et al.: Qualification of a Ni–Cu alloy for the laser powder bed fusion process (LPBF): its microstructure and mechanical properties. Appl. Sci. **10**(10), 3401 (2020)
3. Kimes, K., Myers, K., Klein, A., Ahlfors, M., Stevens, E., Chmielus, M.: Binder Jet 3D printing of 316L stainless steel: effects of HIP on fatigue. Microsc. Microanal. **25**(S2), 2600–2601 (2019)
4. Karakurt, I., Lin, L.: 3D printing technologies: techniques, materials, and post-processing. Curr. Opin. Chem. Eng. **28**, 134–143 (2020)
5. Ulas, M., Aydur, O., Gurgenc, T., Ozel, C.: Surface roughness prediction of machined aluminum alloy with wire electrical discharge machining by different machine learning algorithms. J. Market. Res. **9**(6), 12512–12524 (2020)
6. Xia, C., Pan, Z., Polden, J., Li, H., Xu, Y., Chen, S.: Modelling and prediction of surface roughness in wire arc additive manufacturing using machine learning. J. Intell. Manuf. 1–16 (2021)
7. Zhan, Z., Li, H.: Machine learning based fatigue life prediction with effects of additive manufacturing process parameters for printed SS 316L. Int. J. Fatig. **142**, 105941 (2021)
8. Chen, Z., Wu, X., Tomus, D., Davies, C.H.: Surface roughness of selective laser melted Ti-6Al-4V alloy components. Addit. Manuf. **21**, 91–103 (2018)

Low-Temperature Atmospheric Pressure Plasma for Applications in Flexible and Printed Electronics

Július Vida(✉) and Tomáš Homola

Department of Physical Electronics, Faculty of Science, Masaryk University, Kotlářská 267/2, 611 37 Brno, Czech Republic
{jvida,tomas.homola}@mail.muni.cz

Abstract. Low-temperature plasma processing of surfaces and interfaces is an interesting option for applications in flexible and printed electronics where surface cleaning, activation or functionalization are required. Plasma processing compared to more traditional chemical treatment can be significantly cheaper and faster while eliminating the need for toxic organic solvents. Printed functional coatings often contain organic moieties from solvents, binders and other additives that tune the properties of the precursor liquid. These need to be effectively removed for optimal properties of the resulting film. Plasma can provide reactive and energetic species to the surface of printed films for post-printing processing at significantly lower temperatures compared to conventional thermal annealing. In this paper, we summarize applications of low-temperature atmospheric plasma for cleaning of indium tin oxide (ITO) and fluorine-doped tin oxide (FTO) substrates for improved interface quality between the electrode and the blocking layer in perovskite solar cells. Plasma treatment in order of seconds proved as effective as a chemical treatment for an hour. We also present a way for post-deposition mineralization of mesoporous TiO_2 electron transport layer for perovskite solar cells at temperatures below 70 °C compatible with temperature-sensitive substrates.

Keywords: Low-temperature plasma · Flexible and printed electronics · Perovskite solar cells · DCSBD · Roll-to-roll

1 Introduction

Low-temperature plasma generated at atmospheric pressure has become a versatile tool in a wide range of applications tied to flexible and printed electronics. Unique energetic species generated in plasma, hardly attainable in any other environment, such as electrons, ions, metastable atoms and molecules, radicals and energetic photons interact with the surface of the treated material. Interactions between the plasma species and the surface can have beneficial effects on the treated material if the processing conditions are chosen appropriately.

It has been shown that exposing polymers to oxygen-containing plasma for a very short time leads to significant improvement of the wettability and ink adhesion in the

Y. Wei and S. Chng (Eds.): INCASE 2021, LNME, pp. 66–69, 2022.
https://doi.org/10.1007/978-981-16-5763-4_15

subsequent steps of preparing functional coatings [1]. This is caused by the introduction of new polar functional groups on the surface that enhance surface energy. Plasma can also provide energetic particles for effective removal of organic adsorbates, such as hydrocarbon surface contaminants, from the surface, thus cleaning while maintaining the morphology of the surface mostly intact [2].

Dielectric barrier discharges (DBDs), specifically coplanar DBDs that generate plasma on the surface of a dielectric and with an appropriate geometry of electrodes can produce macroscopically homogeneous diffuse plasma [3] became prominent and efficient tool for surface modification applications. A crucial advantage of plasma generated by DBDs is compatibility with temperature-sensitive flexible materials such as polymeric foils, paper, or textile, which are highly desirable for transition to the roll-to-roll manufacturing, where conventional rigid substrates are hard to incorporate.

Substrates such as glass are still preferred in many applications as most functional coatings require high-temperature post-printing steps for removal of organic additives in the printed film that hinder its properties. Investigations of atmospheric pressure plasma processes such as sintering of metal and metaloxide nanoparticles, conversion of inorganic metal salt-based inks into conductive films and removal of organic moieties from nanoparticle binders by oxidation in plasma can open a way for employment of new substrates, faster processing times at reduced temperature, higher throughput and more environmentally friendly manufacturing technologies [4].

In this contribution, we summarize the effects of low-temperature atmospheric plasma on ITO and FTO surfaces, subsequently employed in perovskite solar cells (PSCs). Rapid plasma treatment processing in order of seconds proved as effective as chemical treatment for an hour. Besides, we show rapid plasma mineralization of mesoporous TiO_2 electron transport layer that can find an application in roll-to-roll manufacture of PSCs.

2 Methods

2.1 Plasma Treatment

Atmospheric-pressure, low-temperature ambient-air plasma was generated by a portable RPS40 plasma source (manufactured by Roplass Ltd., Czech Republic) equipped with a diffuse coplanar surface barrier discharge (DCSBD) plasma unit – a coplanar DBD. The plasma discharge was powered by a continuous-wave sinusoidal AC high voltage at a frequency ~ 19 kHz, while the total power input and volume power density were 40 W and 160 W/cm^3, respectively. Such power density is one of the highest reported for ambient-air plasma generated by DBD. This is related to the unique configuration of the coplanar electrode setup, which enables the generation of a very thin layer of surface plasma with a thickness of approximately 0.1–0.2 mm; therefore no energy losses occur due to the generation of plasma in space where no material treatment is needed. Such "wasted plasma" is very typical for volume DBD, in the course of which plasma is generated in the volume between the plane and cylindrical electrodes and the typical gap in open air is 1–3 mm.

Table 1. Results of fast plasma cleaning of transparent conductive films of ITO and FTO compared to uncleaned and chemically cleaned. Values reproduced with permission from [2] and [6].

	Cleaning method	C:O ratio	WCA
ITO	Uncleaned	1.60	84°
	2-propanol in ultrasonic bath 5 min	0.76	N/A
	5 s plasma treatment	0.19	5°
FTO	Uncleaned	3.89	55°
	Detergent/acetone/IPA in ultrasonic bath 1 h	0.53	35°
	10 s plasma treatment	0.25	7°

3 Results and Discussion

The cleaning of ITO and FTO substrates using rapid atmospheric pressure plasma was compared to standard chemical cleaning procedure. As a figure of merit, the carbon to oxygen (C:O) ratio and water contact angle were chosen and the results are shown in Table 1. Significant decrease in C:O ratio implies successful removal of organic adsorbates in very short time (order of seconds) compared to longer chemical treatment (hours). Decreased water contact angle relates to improved wettability of the surface. *Hvojník et al.* used the plasma-cleaned FTO substrates for preparing high-quality blocking layer of compact TiO_2 for carbon-based perovskite solar cells [6]. They have shown that properly cleaned substrate is crucial for ideal homogeneous layer with high quality and plasma provides a way to clean the substrate in a matter of seconds.

Preparing the mesoporous TiO_2 (m-TiO_2) electron transport layer (ETL) for perovskite solar cells (PSCs) at low temperature is a step forward towards flexible photovoltaics. *Homola et al.* used a ink consisting of TiO_2 nanoparticles dispersed within organosilica binder [5]. The methyl groups in the binder hinder the electronic properties of the resulting ETL and the as-deposited film is inappropriate for PSCs. However, after

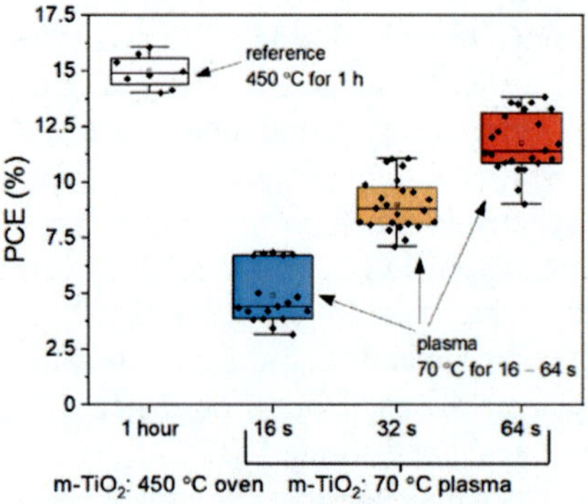

Fig. 1. Power conversion efficiency of perovskite solar with plasma-processed electron-transport layer compared to reference high-temperature process. Reprinted with permission from [5].

just one minute of ambient air plasma treatment, the methyl groups from the binder were successfully oxidized and the properties of the film improved significantly. This allowed quality deposition of the perovskite and resulted in power conversion efficiency (PCE) of 12 ± 1% compared to 15 ± 1% for the reference high-temperature process (see Fig. 1).

4 Conclusion

We demonstrated several examples of low-temperature plasma at atmospheric pressure for surface modifications for applications in flexible and printed electronics. Plasma-cleaning was demonstrated on conductive FTO and ITO substrates. FTO substrates were then successfully employed in carbon-based perovskite solar cells. Low-temperature process for m-TiO_2 ETL in perovskite solar cells was compared to high-temperature annealing. One minute of ambient air plasma treatment successfully removed organic moieties from organosilica binder and the achieved PCE was within 80% of conventional thermal route.

Acknowledgements. This research was supported by GACR project 19-14770Y funded by Czech Science Foundation.

References

1. Homola, T., Wu, L.Y.L., Černák, M.: J. Adhes. **90**(4), 296–309 (2014). https://doi.org/10.1080/00218464.2013.794110
2. Homola, T., et al.: Appl. Surf. Sci. **258**(18), 7135–7139 (2012). https://doi.org/10.1016/j.apsusc.2012.03.188
3. Homola, T., Matoušek, J., Kormunda, M., Wu, L.Y.L., Černák, M.: Plasma Chem. Plasma Process. **33**(5), 881–894 (2013). https://doi.org/10.1007/s11090-013-9467-3
4. Sui, Y., Zorman, C.A., Sankaran, R.M.: Plasma Process. Polym. (January), 1–25 (2020). https://doi.org/10.1002/ppap.202000009
5. Homola, T., et al.: ACS Appl. Energy Mater. **3**(12), 12009–12018 (2020). https://doi.org/10.1021/acsaem.0c02144
6. Hvojnik, M., et al.: Gemeiner. Mater. Sci. Semicond. Process. **131**(April) **(2021)**. https://doi.org/10.1016/j.mssp.2021.105850

Preparation for Metallic Composite Material with High Surface Strength-Corrosion Resistance

Baisong Cheng, Fengxia Wei, Gary Wei Hock Teh, Augustine Kok Heng Cheong, Coryl Jing Jun Lee, Junyi Liu, and Dennis Cheng Cheh Tan(✉)

Institute of Materials Research and Engineering (IMRE), A*STAR (Agency for Science, Technology and Research), 2 Fusionopolis Way, Singapore 138634, Singapore
{Cheng_Baisong,dennis-tan}@imre.a-star.edu.sg

Abstract. Hot-compression bonding combined with high-temperature heat treatment were applied to fabricate a metallic composite material, which consisted two surface layers made of conventional 316L stainless steel and a matrix made of additively manufactured (AM) high-nitrogen steel. The results showed that the dual action of large compressive deformation and high-temperature heat treatment promoted the dynamic recrystallization of grains in the bonding area and finally realized its full metallurgical bonding. The diffusion of nitrogen atom from matrix to surface improves both the hardness and corrosion resistance of the surface layer. At the same time, the dense surface layer compensates the corrosion resistance reduction caused by pores of AM high-nitrogen steel. In summary, this method provided a new direction for optimizing surface performance for AM metal parts.

Keywords: Metallic composite material · Hot-compression bonding · High-temperature heat treatment · Surface strengthening · Interface

1 Introduction

Pores are the main defect of additively manufactured (AM) metal parts [1–3], which would severely deteriorates the mechanical property and the corrosion resistance. At present, the principal measures for decreasing the porosity include the printing parameters optimization (laser power, hatch spacing, layer thickness, scan speeds, etc.) and powder parameters optimization such as size distribution, degree of sphericality, purity, and mixing uniformity, etc. However, these optimization measures are time consuming. As we know, surface treatment not only can meet the usage requirements, but also save the cost of material manufacturing. For example, the well-known thermal spraying [4] and cold spraying [5] have been widely used in industry. However, among various spraying techniques, there is still an issue of poor adhesion, as well as the same porosity problem, especially for the printed parts with more surface pores. For many metal parts fabricated by AM, the application of spraying techniques has limitations. Therefore, in this work, we deployed hot-compression bonding [6] combined with high-temperature heat treatment to improve the surface performance for AM metal parts.

Y. Wei and S. Chng (Eds.): INCASE 2021, LNME, pp. 70–73, 2022.
https://doi.org/10.1007/978-981-16-5763-4_16

2 Experimental Method

2.1 Hot-Compressing Bonding

A compression machine equipped with high temperature furnace was used to do the thermal compression experiment. The schematic of hot-compression bonding process is shown as Fig. 1. The three-layered sample is composed of two layers of 316L stainless steels and one layer of high-nitrogen steel in the middle. The heating curve from room temperature to 1150 °C is synchronized with the compression strain from 0 to 0.5. When the compression strain reaches 0.5 and the heating temperature reaches 1150 °C, the condition holds for 10 min followed by air cooling.

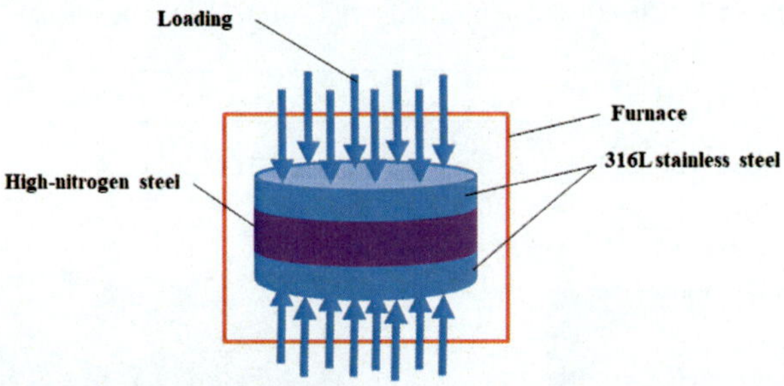

Fig. 1. The schematic of hot-compression bonding process

2.2 High-Temperature Heat Treatment

After hot-compression bonding, the sample was then treated according to the heating curve as shown in Fig. 2. The main purpose of heat treatment is to promote the metallurgical combination and dynamic recrystallization of grains in the bonding area between 316L stainless steel and high-nitrogen steel.

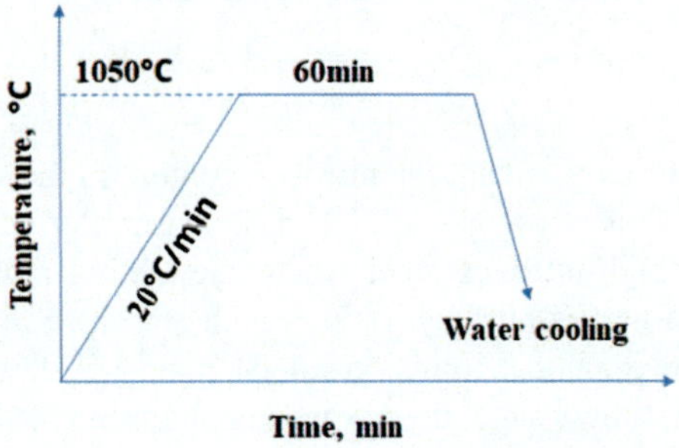

Fig. 2. The diagram of heat treatment process

3 Result and Discussion

3.1 Chemical Composition and Porosity

Before hot-compression bonding, the chemical composition of 316L stainless steel made by conventional technology (smelting, casting, forging, rolling and solid solution treatment) is shown in Table 1. Besides nitrogen-free, it can also be assumed that there are no pores in 316L stainless steel plates. The porosity of high-nitrogen steel fabricated by selective laser melting (SLM) was examined with optical microscopy (OM), as shown in Fig. 3. A large number of pores are clearly visible, ranging in size from a few microns to tens of microns, with a maximum size 101.33 μm, and the porosity is around 2.62% .

Table 1. Chemical composition (wt%) of 316L stainless steel

C	Si	Mn	P	S	Cr	Ni	Mo
0.023	0.320	1.690	0.041	0.027	16.720	10.070	2.060

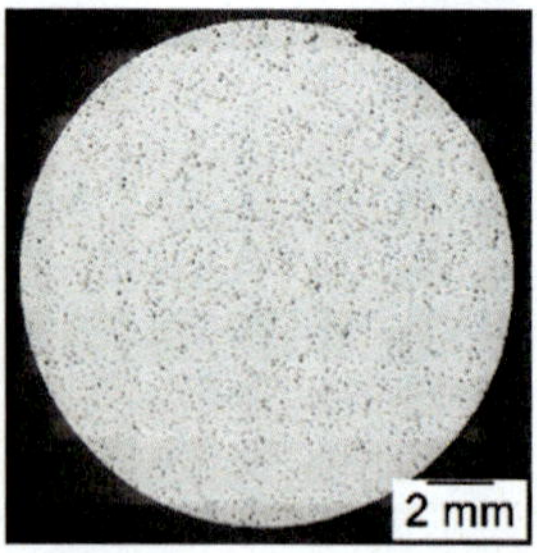

Fig. 3. Porosity of high-nitrogen steel

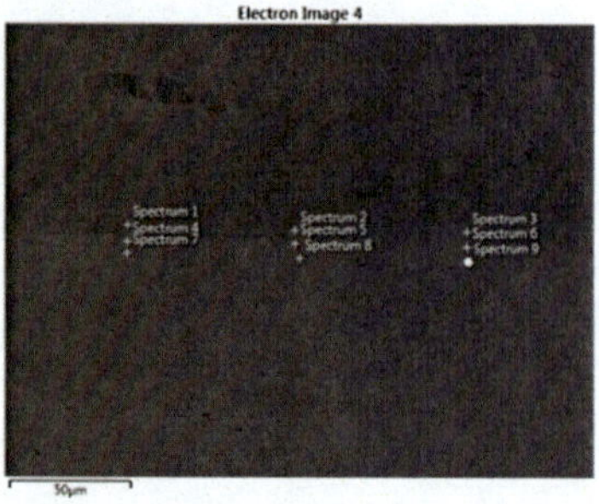

Fig. 4. Nitrogen content testing by EDX

Table 2. Distribution of nitrogen content (wt%)

P_1	P_2	P_3	P_4	P_5	P_6	P_7	P_8	P_9
0.7	0.8	0.6	0.5	0.7	0.3	0.3	0.2	0

After high-temperature treatment, the nitrogen content in the sample was tested by Energy-dispersive X-ray spectroscopy (EDX), as shown in Fig. 4. The top half (darker part) of the picture is high-nitrogen steel where the testing points (P1, P2 and P3) are located and contain higher nitrogen ratio, which are close to the original content (around 0.87wt%). However, the points (from P4 to P9) in lower half (light color) which belong to 316L stainless steel, the nitrogen contents are less. Therefore, nitrogen atoms have partially spread into the surface from matrix. The farther away from bonding interface, the lower the nitrogen content. The diffused interface between layers proves good metallurgical combination. The large deformation and high-temperature promote the dynamic recrystallization of grains near the interface.

3.2 Micro-hardness and Corrosion-Resistance Analysis

Micro-hardness tests of sample surface were carried out with a Vickers hardness tester. The load was 500 g and dwell time was 10 s. Under solid solution strengthening, the average value of hardness reaches about 182.15HV.

Finally, reliable results [7, 8] have been obtained on the beneficial effect of nitrogen on corrosion resistance. Particularly, nitrogen can significantly improve pitting and crevice corrosion of austenitic stainless steel. As a result, the sample prepared by this method will also have high corrosion resistance.

4 Conclusion

1) The dual action of large compressive deformation and high-temperature heat treatment promotes the dynamic recrystallization of grains in the bonding area and metallurgical bonding. This method can be used to fabricate metallic composites with high strength and corrosion resistance.
2) The partial diffusion of nitrogen element from matrix to surface layer can improve the hardness of the surface layer, reaching around 182.15HV.
3) Nitrogen can improve the corrosion resistance of 316L stainless steel, and the dense surface layer can compensate the lack of corrosion resistance caused by pores of additively manufactured high-nitrogen steel.

References

1. Sola, A., Nouri, A.: Microstructural porosity in additive manufacturing: the formation and detection of pores in metal parts fabricated by powder bed fusion. J. Adv. Manuf. Process. 1–21 (2019)
2. Slotwinski, J.A., Garboczi, E.J., Hebenstreit, K.M.: Porosity measurements and analysis for metal additive manufacturing process control. J. Res. Natl. Inst. Stand. Technol. **119**, 494–528 (2014)
3. Martin, A.A., et al.: Dynamics of pore formation during laser powder bed fusion additive manufacturing. Nat. Commun. **10**, 1–10 (2019)
4. Yin, S., Cizek, J., Suo, X., Li, W., Liao, H.: Thermal spray technology. Adv. Mater. Sci. Eng. **2019**, 1–2 (2019)
5. Yin, S., Chen, C., Suo, X., Lupoi, R.: Cold-sprayed metal coatings with nanostructure. Adv. Mater. Sci. Eng. **2018**, 1–19 (2018)
6. Yang, X., Li, W., Feng, Y., Siqi, Y., Xiao, B.: Physical simulation of interfacial microstructure evolution for hot compression bonding behavior in linear friction welded joints of GH4169 superalloy. Mater. Des. **104**, 436–452 (2016)
7. López, D., Falleiros, N.A., Tschiptschin, A.P.: Effect of nitrogen on the corrosion-erosion synergism in an austenitic stainless steel. Tribol. Int. **44**, 610–616 (2011)
8. Bayoumi, F.M., Ghanem, W.A.: Effect of nitrogen on the corrosion behavior of austenitic stainless steel in chloride solutions. Mater. Lett. **59**, 3311–3314 (2005)

Surface and Subsurface Characterisation

Novel Interpretation of Volta Potential Measurements from *In-Situ* SKPFM for Localised Corrosion Analysis in Aluminium Alloys

YanHan Liew and Daniel J. Blackwood(✉)

Department of Materials Science and Engineering, National University of Singapore, Singapore, Singapore
{YHLIEW,msedjb}@nus.edu.sg

Abstract. Volta potential measurements from Scanning Kelvin Probe Force Microscopy (SKPFM) have been used over the years to discuss the nobility of an alloy's surface and its tendency to localised corrosion. This has led to classical interpretations that may (unfortunately) over-generalise the actual Volta potential behaviour of the surface during ongoing localised corrosion. Consequently, issues on correlating Volta potential measurements to corrosion potentials and mechanisms arise. With the advent of *in-situ* SKPFM, which allows for the Volta potential measurements during ongoing dynamic surface electrochemical and/or physicochemical reactions, novel observations regarding Volta potential evolution are available. This calls for an updated and re-evaluated interpretation of the Volta potential measurements from SKPFM relating to localised corrosion analysis. The novel interpretation includes an additional layer of information, as compared to classical interpretations; there is greater information on the role of interfacial parameters and time (kinetics). Optimistically, this will bridge the gap of Volta potential measurements and its correlation to corrosion potentials.

Keywords: Localised corrosion · SKPFM · Aluminium alloys · Humidity

1 Introduction

Analysis of an alloy surface's Volta potential [1] can allow for the identification of regions in the microstructure that have a higher or lower propensity to take part in electrochemical reactions, and therefore providing information on possible localised corrosion initiation and/or propagation sites. Traditionally, researchers have concluded that upon exposure to an electrolyte, intermetallic particles (IMPs) in aluminium alloys (AA) causes micro-galvanic interactions with their surroundings [2], be it the peripheral matrix or even other IMPs; these microgalvanic interactions can then lead to local dissolution of the peripheral matrix or the IMP itself, forming localised corrosion sites.

The Scanning Kelvin Probe Force Microscopy (SKPFM) provides, with remarkable spatial resolution (nm-scale) and energy resolution, the characterisation of the Volta

Y. Wei and S. Chng (Eds.): INCASE 2021, LNME, pp. 77–82, 2022.
https://doi.org/10.1007/978-981-16-5763-4_17

potential difference between the AA surface and the probe tip [3]. This has produced some successes in attempts to correlate Volta potential measurements with the corrosion potential of pure metals [4]. However, the correlation cannot be generalised and is only applicable to certain systems [3]. A possible reason for the lack of applicability of the correlation is postulated to be due to the "missing" electrolyte during SKPFM [1], as measurements were done *ex-situ* in (dry) air.

Therefore, to overcome this issue, an *in-situ* SKPFM, which incorporates a relative humidity variation to provide for the surface electrolyte, was recently developed [5, 6] and validated [7–9]. This has opened up novel interpretations of the Volta potential measurements from SKPFM, and its relationship to localised corrosion mechanisms. The classical and possible novel interpretations of Volta potential measurements from SKPFM will be the focus of our discussion in this paper.

2 Methods

The *in-situ* SKPFM setup basically involves incorporating a humidifier to the Bruker Dimension Icon atomic force microscope (AFM), together with the Bruker cell setup originally designed for nitrogen gas purging (Fig. 1). SKPFM measurements were taken together with the variation of humidity from dry (<30% RH) to wet (>80% RH). Therefore, the SKPFM measurements were considered to be done *in-situ*, during the ongoing electrochemical and/or physicochemical surface reactions.

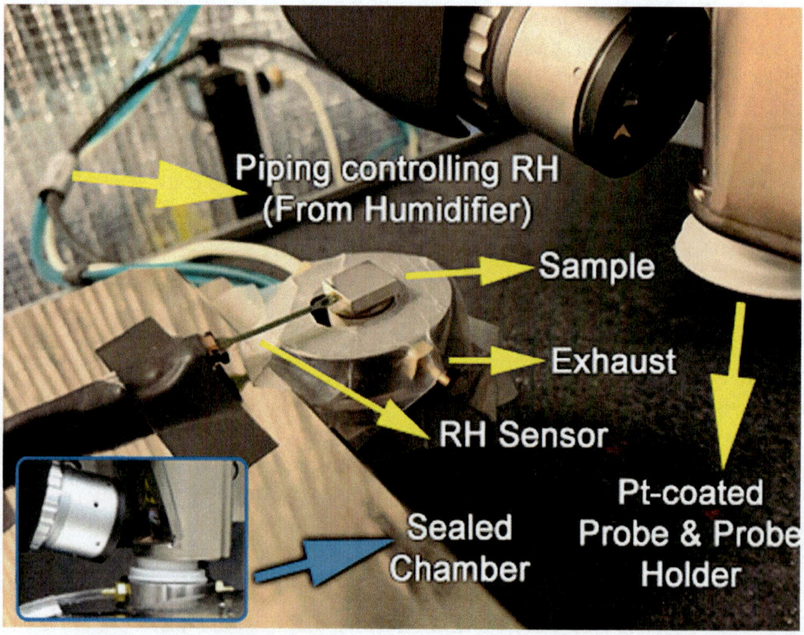

Fig. 1. Schematic of the controlled sealed environment setup to vary the relative humidity from 25% to 85% during the *in-situ* SKPFM experiments. The inset illustrates the sealed chamber during the SKPFM scans.

For further details regarding the *in-situ* SKPFM setup, readers are referred to the authors' previous works and other literature [5–9].

3 Discussions

3.1 Classical Interpretation of Volta Potential Measurements (from *Ex-Situ* SKPFM)

Intrinsically, the Volta potential of a metal is related to the energy required to bring an electron from a point infinitely far away from the surface to a point just outside the metal's surface. However, it should be noted that Volta potential measurements of the metal's surface from SKPFM involves a reference point (probe tip) and is therefore a relative value. Nevertheless, proper calibration of the Volta potential of the probe tip (for example, against a reference sample [10]) will still allow for valuable Volta potential measurements to be analysed.

Importantly, the relative Volta potential differences, especially between that of the IMPs and its surrounding matrix, has been described as the driving force for micro-galvanic corrosion cells [1]. Traditional *ex-situ* SKPFM measurements done in (dry) air after immersion in electrolyte have assisted in elucidating surface reactions over an extended period of electrolyte immersion time, such as the preferential oxidation of the peripheral matrix surrounding an IMP with a positive Volta potential (vs matrix) or the dissolution of certain elements in the IMP together with their re-deposition [11].

The multitude of literature using the *ex-situ* SKPFM interpretations have instilled a generalisation (or even assumption) that the *ex-situ* measured regions with a more positive (more noble) Volta potential are likely the cathodes during localised corrosion, while the regions with a less positive (more active) Volta potential are likely the anodes; this is even though the localised corrosion process being described was not physically occurring during the time period the *ex-situ* SKPFM scans in (dry) air were collected. Hence, this generalisation is an over-simplification of the actual ongoing surface reactions that occur during the localised corrosion itself; it has been proven that the aforementioned classical interpretations cannot be generalised for all systems [3].

While these classical interpretations have provided insights, they do not provide the entire picture; especially on Volta potential evolution during the ongoing corrosion reactions, when the samples were immersed in the electrolyte. Figures 2a and 2d, measured in dry air (low RH $< 30\%$), can illustrate the snap shots used for analysis in classical *ex-situ* SKPFM measurements; these are missing important changes during ongoing reactions with the electrolyte (Figs. 2b and 2c). Hence, it can be described that *ex-situ* SKPFM measurements in (dry) air, and their corresponding classical interpretations, are based on understanding the state of the surface after the physicochemical and/or electrochemical reactions involved have been heavily limited or halted.

3.2 Novel Interpretation of Volta Potential Measurements (from *In-Situ* SKPFM)

With the added involvement of elevated relative humidity ($>80\%$ RH), a (pseudo thin-film) form of electrolyte is present on the alloy surface during the SKPFM scans;

it has been validated that sufficient electrolyte is present for surface electrochemical and/or physicochemical reactions to occur [7–9]. These Volta potential measurements are considered to be collected *in-situ*, during dynamic ongoing localised corrosion processes.

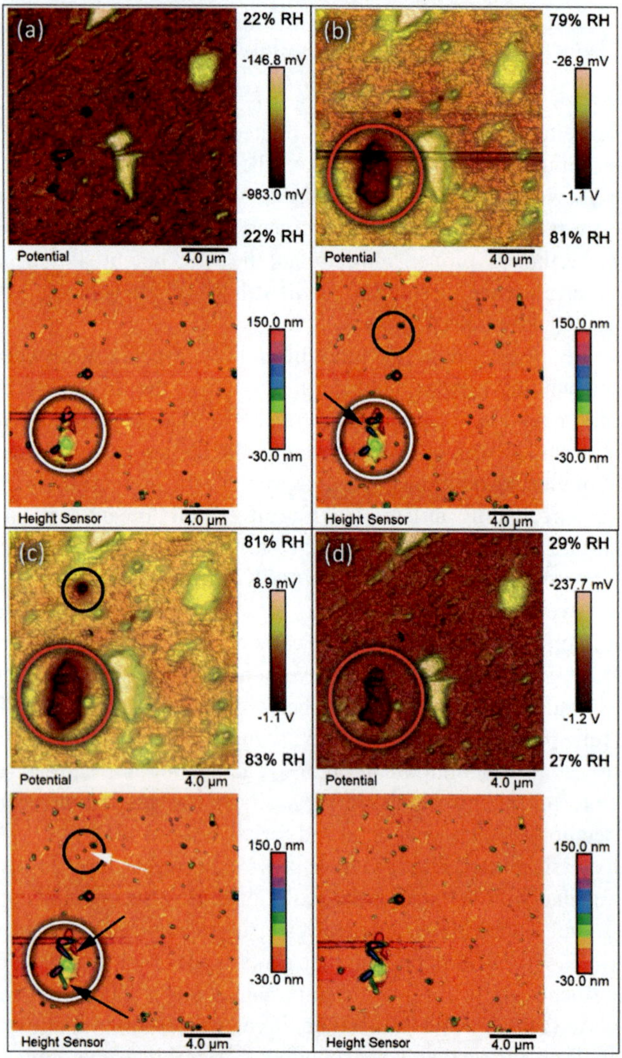

Fig. 2. Volta potential (top) and topography (bottom) maps of the sensitized AA5083 microstructure acquired during one of the *in-situ* SKPFM scan cycles. 2(a) and 2(d) are taken at low humidities ($<30\%$ RH) and can be likened to *ex-situ* SKPFM snap shots used for classical interpretations. 2(b) and 2(c) shows the abrupt and significant Volta potential changes of the surface when exposed to high humidities (ca. $>80\%$ RH) at the appropriate time; topography changes coincided with the Volta potential changes. Important changes during the ongoing corrosion reactions, 2(b) and 2(c), would not have been recorded during classical *ex-situ* SKPFM. Reproduced from [8].

However, with an electrolyte on the surface, the interpretation of Volta potential measurements from SKPFM has to be updated and re-evaluated. Volta potential differences at the electrode/electrolyte interface has to be considered [12], implying that interfacial parameters play a greater role now. Overall, the Volta potential measurements involve a convoluted summation of all the interfaces that are within the tip-sample interaction; these interfaces can be porous or exhibit different properties when exposed to the electrolyte. Hence, *ex-situ* measurements without the presence of the electrolyte are not sufficient to correlate to corrosion mechanisms.

Furthermore, this means that it is necessary to inculcate novel interpretations of the Volta potential measurements from SKPFM, whereby more positive (more noble) Volta potential measured for a region does not necessarily mean that it will remain noble when exposed to the corrosive environment [7, 8]. Localised corrosion has a dynamic nature, and it has been shown that previously noble (more positive Volta potential vs matrix) regions in the alloy can exhibit an abrupt and severe nobility loss (>400 mV) when exposed to an electrolyte at the appropriate time [8] (red circles in Fig. 2, with corresponding topography changes illustrated by white circles). Abrupt and severe nobility losses measured *in-situ* can indicate the dissolution of active elements with loss of electrons, producing an electropositive surface at that point in time during the measurement; time (and kinetics) also plays a greater role now.

Transient and momentary nobility losses of a localised region have also been seen to occur only during high humidities [8] (black circles in Fig. 2), when the electrolyte is suitable for electrochemical and/or physicochemical reactions, indicating that *ex-situ* measurements can miss out on important Volta potential measurements.

Therefore, an additional layer of information, regarding the involvement of the electrolyte, is included during *in-situ* SKPFM scans; this information includes corrosion kinetics, and might even involve the ionic/current flows in the electrolyte during localised corrosion [9]. Optimistically, together with classical interpretations, Volta potential measurements from SKPFM (both *in-situ* and *ex-situ*) can work towards bridging the relationship of Volta potential measurements with corrosion potentials.

4 Conclusions

- Interpretation of Volta potential measurements has to be updated and re-evaluated to avoid over-generalisation of classical interpretations from *ex-situ* SKPFM.
- Novel interpretations from *in-situ* SKPFM includes greater information on the kinetics of localised corrosion, and possibly even information on the ionic/current flows in the electrolyte.

References

1. Örnek, C., Leygraf, C., Pan, J.: On the Volta potential measured by SKPFM–fundamental and practical aspects with relevance to corrosion science. Corros. Eng. Sci. Technol. **54**(3), 185–198 (2019)

2. Birbilis, N., Buchheit, R.: Electrochemical characteristics of intermetallic phases in aluminum alloys. J. Electrochem. Soc. **152**(4), B140–B151 (2005)
3. Rohwerder, M., Turcu, F.: High-resolution Kelvin probe microscopy in corrosion science: scanning Kelvin probe force microscopy (SKPFM) versus classical scanning Kelvin probe (SKP). Electrochim. Acta. **53**(2), 290–299 (2007)
4. Schmutz, P., Frankel, G.: Characterization of AA2024-T3 by scanning Kelvin probe force microscopy. J. Electrochem. Soc. **145**(7), 2285 (1998)
5. Örnek, C., Engelberg, D.: SKPFM measured Volta potential correlated with strain localisation in microstructure to understand corrosion susceptibility of cold-rolled grade 2205 duplex stainless steel. Corros. Sci. **99**, 164–171 (2015)
6. Örnek, C., Liu, M., Pan, J., Jin, Y., Leygraf, C.: Volta potential evolution of intermetallics in aluminum alloy microstructure under thin aqueous adlayers: a combined DFT and experimental study. Top. Catal. **61**, 1169–1182 (2018)
7. Örnek, C., Leygraf, C., Pan, J.: Real-time corrosion monitoring of aluminum alloy using scanning Kelvin probe force microscopy. J. Electrochem. Soc. **167**(8), 081502 (2020)
8. Liew, Y., Örnek, C., Pan, J., Thierry, D., Wijesinghe, S., Blackwood, D.: *In-situ* time-lapse SKPFM investigation of sensitized AA5083 aluminum alloy to understand localized corrosion. J. Electrochem. Soc. **167**(14), 141502 (2020)
9. Liew, Y.: Impacts of precipitates and electrochemical aspects of localised corrosion in industrial alloys. National University of Singapore (2021)
10. Efaw, C., da Silva, T., Davis, P., Li, L., Graugnard, E., Hurley, M.: Toward improving ambient Volta potential measurements with SKPFM for corrosion studies. J. Electrochem. Soc. **166**(11), C3018–C3027 (2019)
11. Zhu, Y., et al.: Micro- and nano-scale intermetallic phases in AA2070-T8 and their corrosion behavior. Electrochim. Acta. **319**, 634–648 (2019)
12. Trasatti, S.: The absolute electrode potential: an explanatory note (recommendations 1986). Pure Appl. Chem. **58**(7), 955–966 (1986)

Minerals Disorder Determination in Special Cases of Metastatic Bone Due to Breast Cancer Using LIBS Technique

Elshaimaa M. Emara[1,2](✉), Bing Wang[1,2], Haiying Song[1,2], and Shibing Liu[1,2](✉)

[1] Strong-Field and Ultrafast Photonics Lab, Faculty of Materials and Manufacturing, Beijing University of Technology, Beijing 100124, China

[2] Key Laboratory of Trans-Scale Laser Manufacturing Technology, Ministry of Education, Beijing University of Technology, Beijing 100124, China

Abstract. Minerals are essential elements in the human body; they are involved in many vital biological functions; such as calcium and phosphate which are involved in the muscle and bone formation. The blood serum is the monitor of the body content, so determining calcium and phosphate level in the blood serum is so important to work as a biomarker for detecting some diseases and foremost cancers. The breast cancer is one of the most life threatening disease for women around the world and if is followed by bone metastasis, so the survivability is so limited. Rapid detection and determination of calcium and phosphate disorder has been occurred in the patients could lead to rescue many of them, so using LIBS in this process can achieve it in the proper way in terms of time saving and low cost.

Keywords: LIBS · Breast cancer · Serum · Calcium · Phosphate · Bone metastasis

1 Introduction

Calcium and phosphate are very important for the human body, for both muscles and skeleton. They can be involved in the bone abnormalities if there is disorder in their serum concentrations [1]. The blood content of calcium and phosphate is coming mainly from bone [2, 3]. Calcium is an essential element for numerous biochemical processes in the body; it is considered as main element in the bone growth and development [4].

The serum calcium level can be highly abnormal in case of malignancies or in the parathyroid gland disorder [5]. In the presence of malignancy, the serum calcium level could be associated with hypophosphatemia in case of breast cancer; it's called severe hypercalcemia which points to bone metastasis [6]. Breast cancer could lead in many cases (20–30%) to osteolytic bone metastases which certainly causes hypercalcemia [7].

Phosphate is one of the most important and copious mineral in the body. Adjusting the phosphate level helps keeping the body bones normal and healthy [8]. The reduction in phosphate is in complication with hypercalcemia, especially which associated with cancer.

Y. Wei and S. Chng (Eds.): INCASE 2021, LNME, pp. 83–87, 2022.
https://doi.org/10.1007/978-981-16-5763-4_18

Dehydration causes phosphate disorder as to make what called hypophosphatemia, all of these etiologies involve in the problem of hypercalcemia [9].

It has been noticed that more than 30% of women with breast cancer may develop bone metastasis especially in the skeleton [10].

Some studies have elucidated that the serum levels of calcium and phosphate in patients with cancers specially breast cancer have been altered [11, 12] and leading to disorders in these minerals which are involved in osteolytic bone metastasis [10] Laser-Induced Breakdown Spectroscopy (LIBS) is a spectroscopic technique to achieve the elemental analysis for various materials' forms; solid, liquid and gas [13]. So, this study shows the use of LIBS in the detection of the cases of severe hypophosphatemia associated with severe hypercalcemia to discover as a primary investigation of breast cancer the presence of bone metastasis through serum minerals.

2 Material and Method

2.1 Serum Samples

Blood samples have been voluntary collected by hematology expert from breast cancer patients and some other healthy women, all were in the age range 45–60 yrs old. The samples were divided into two groups, one for the healthy controls (8 women) and the other for breast cancer patients (8 women samples). The preparations have been done in a sanitizing area of laminar flow cabinet. A drop of serum was put on glass slide for each sample, and then let dry under the sanitizing cabinet. After drying, the samples were ready for processing by LIBS.

2.2 LIBS Setup

Nd:YAG pulsed laser "Continuum; Surelite II" was used, it was operating at its fundamental wavelength (1064 nm) at repetition rate of 10 Hz, the pulse energy was 25 mJ and the pulse width was 10 ns, a 100 mm focusing lens was used to focus the laser beam on the target. A schematic diagram in Fig. 1 shows the LIBS setup that has been used for this study. Collective lens and the optical fiber were used to transfer the spectra to the spectrometer "Andor, Mechelle 5000" with spectral resolution of $\lambda/\Delta\lambda = 5000$, it was coupled with "DH334" ICCD camera. This setup was controlled by the "Stanford, DG645" delay generator and automatized by licensed software "Andor SOLIS 64-bit".

2.3 LIBS Parameters

To produce a good LIBS signal with respect to signal to noise ratio (SNR). Gate time delay of 300 ns with a gate time width of 1000 ns have been chosen and a set of 30 spectra has been collected for each sample. The calculations and manipulations of the collected spectra were performed using OriginPro8 software to measure the qualitative and quantitative analysis of the elemental composition in the serum samples.

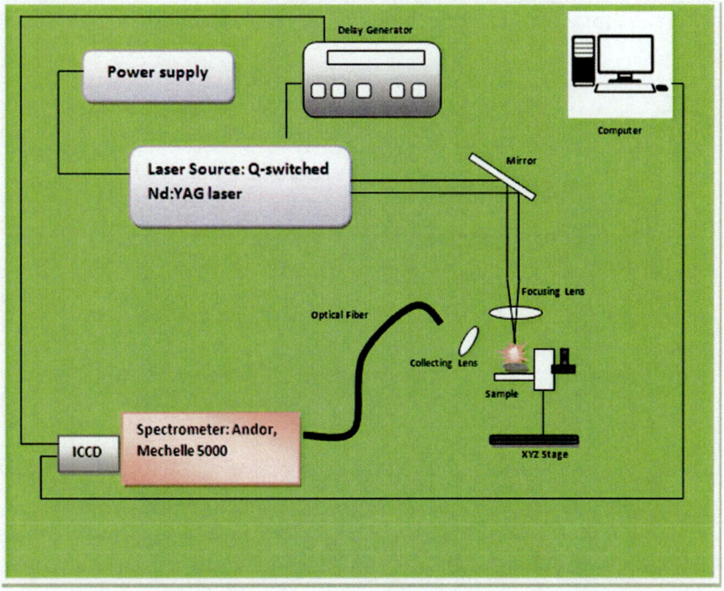

Fig. 1. Schematic diagram of the experimental LIBS setup.

3 Results and Discussion

Most of clinical tests are performed to quantify the concentration of various constituents of the liquid portion of blood (serum) [14], so this kind of tests became very important in the cancer diagnosis.

This study was conducted to elucidate the presence of phosphate disorder (hypophosphatemia) and calcium disorder (hypercalcemia) and finding the relation between them in the cases of breast cancer which cause bone metastasis. The spectral data of LIBS have been collected under the hypothesis of local thermal equilibrium (LTE) from different samples; healthy and breast cancer groups.

Table 1 shows that the serum phosphate levels in the control and breast cancer samples were 69.47 and 17.63 respectively. For illustration; Fig. 2-a can show the huge reduction in P concentration in case of breast cancer below that in the healthy group. By applying statistical test (*t*-test) for these results, it was obvious that there is a significant difference between the two studied groups within $p < 0.05$. The patients in the breast cancer group suffer from severe hypophosphatemia; that is related to the occurrence of bone metastasis associated with breast cancer.

In the contrary, Ca level is observed very high in the breast cancer group by a great percentage than that in control group as shown in Fig. 2-b. The serum calcium level was 203.16 and 570.56 in healthy controls and breast cancer respectively and great significant difference has been found between the two groups with $p < 0.05$ in the statistical t-test. Most malignancy cases suffer from hypercalcemia [15], it is obvious here from the results there was severe hypercalcemia in the patients group. These results are in compliance with the study of Liou JH. et al.; 2006, who showed that there is strong correlation between hypercalcemia and different kinds of cancers specially female reproductive organs [16].

Table 1. The statistical data with t-test for phosphate and calcium in controls and breast cancer

Element level		Average	Standard deviation
P level	Control	69.47	1.48
	Breast	17.63	1.19
Ca level	Control	203.16	7.02
	Breast	570.56	9.11

$P < 0.05$

The percentages of phosphate and calcium respectively in the studied breast cancer samples were about 25% and 280% of the healthy group and these are very abnormal results even with the presence of breast cancer but because in this case there was bone metastasis which was examined clinically for those patients, a great shortage of phosphate and a huge increase of calcium have been discovered. These findings give the LIBS technique a new ability to distinguish between normal samples, breast cancer samples and breast cancer-bone metastasis samples.

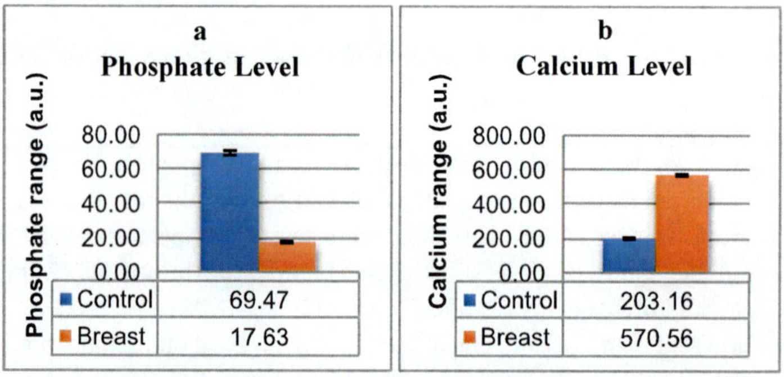

Fig. 2. Relative concentrations of the serum level from healthy control and breast cancer of (a) phosphorus and (b) Calcium

4 Conclusion

This study was conducted to show the abilities of LIBS technique to detect trace elements such as phosphate and calcium in the serum samples.

The presence of hypophosphatemia and hypercalcemia problems have been investigated subsequent breast cancer in case of bone metastasis. The significant difference value of $p = 1.07\text{E}{-18} < 0.05$ between control group and malignancy group in phosphate level was calculated. There is also significant difference between control group and cancer patients group in serum calcium level where $p = 1.4E-19 < 0.05$. The results indicate a continuous metastasis of bone in the presence of breast cancer and severe hypophosphatemia and hypercalcemia. The results disclosed the relation between metastatic breast cancer and the minerals disorders. This study confirmed the prominence and impact of using LIBS technique in investigation of biological problems especially the cases of metastatic cancers.

References

1. Shaker, J.L., Deftos, L.: Calcium and phosphate homeostasis. In: De Groot, L.J., Chrousos, G., Dungan, K. et al. (eds.) Endotext [Internet]. MDText.com, Inc., South Dartmouth (2000)
2. Drezner, M.K.: Phosphorus homeostasis and related disorders. In: Bilezikian, J.P., Raisz, L.G., Rodan, G.A. (eds.) Principles of Bone Biology, 2nd edn (2002)
3. Yanagawa, N., Lee, D.B.N.: Renal handling of calcium and phosphorus. In: Coe, F.L., Favus, M.J. (eds.) Disorders of Bone and Mineral Metabolism (1992)
4. Bronner, F.: Mechanisms of intestinal calcium absorption. J. Cell. Biochem. **88**(2), 387–393 (2003)
5. Russo, R.E.: Laser ablation in analytical chemistry - a review. Talanta **57**(3), 425–451 (2002)
6. Shepard, M.M.: Hypercalcemia. Am. J. Med. Sci. **334**(5), 381–385 (2007)
7. Marks, J.: Phosphate homeostasis and the renal-gastrointestinal axis. Am. J. Physiol. Renal Physiol. **299**, 285–296 (2010)
8. Andrew, F.: Hypercalcemia associated with cancer. N. Engl. J. Med. **352**, 373–379 (2005)
9. Scheid, V.: Clinical course of breast cancer patients with osseous metastasis treated with combination chemotherapy. Cancer **58**(12), 2589–2593 (1986)
10. Halabay, R.: Acid phosphatase in human breast tumors. Breast Cancer Res **53**(2), 301–305 (1984)
11. Solimando, D.A.: Overview of hypercalcemia of malignancy. Am. J. Health-Syst. Pharm. **58**(3), 4–7 (2001)
12. Ohsako, T.: Two cases of stage IV breast cancer with severe hypercalcemia. Gan To Kagaku Ryoho **33**(2), 223–266 (2006)
13. Cremers, D.A.: Direct detection of beryllium on filters using the laser spark. Appl. Spectrosc. **39**, 57–63 (1985)
14. Cremers, D.A., Radziemski, L.J.: Handbook of Laser-Induced Breakdown Spectroscopy. Wiley, Chichester (2006)
15. Sommer, S.: The phosphatonins and the regulation of phosphate transport and vitamin D metabolism. J Steroid Biochem. Mol. Biol. **103**(3–5), 497–503 (2007)
16. Liou, J.H.: Paraneoplastic hypercalcemia with metastatic calcification-clinicopathologic studies. Kaohsiung J. Med. Sci. **22**, 85–88 (2006)

Correlation Study Between Barkhausen Noise and Depth-Resolved Residual Stress Upon Shot Peening

Wei Zhen Iris Koh[1] and Yuefan Wei[2(✉)]

[1] Nanyang Technological University, 50 Nanyang Ave, Singapore 639798, Singapore
KOHW0073@e.ntu.edu.sg

[2] Advanced Remanufacturing and Technology Centre, A*STAR, 3 Cleantech Loop, #01/01 CleanTech Two, Singapore 637143, Singapore
wei_yuefan@artc.a-star.edu.sg

Abstract. This research work examined the use of non-destructive Barkhausen noise technique in establishing residual stress (RS) depth profile in shot-peened specimens. The main objective was to produce a quantitative study of various Barkhausen noise parameters, including root-mean-square (RMS), full-width at half maximum (FWHM), peak average and pulse counts. A correlation was developed between Barkhausen noise parameters with residual stress values that were measured using X-ray diffraction technique. It was noticed that Barkhausen noise generated was highest at shallower depth of 50 μm and lowest at deeper depth of 400 μm, i.e. Barkhausen noise decreased with increasing depth for all the samples. This was attributed to the shot peening treatment being more effective on the near surface region of the material than in its bulk. After a thorough analysis of the key Barkhausen parameters, RMS was determined to produce the best correlation to RS. The depth profiles of RMS and RS were inversely proportional, and this trend was best reflected in shallower depths of 50 μm to 250 μm where the shot-peening process had greater impact on the specimens.

Keywords: Residual stress · Barkhausen noise · Depth profile · Shot peening

1 Introduction

Residual stress (RS) arises from the transformation of the shape or property of a material after its post-processing, and is retained in the material body at equilibrium with its surroundings. Compressive RS delays crack initiation and retards crack growth, which thereby enhances the fatigue properties and performance of components [1]. Compressive RS can be induced in the near-surface (less than 1mm from free surface) of a component by various peening processes such as shot peening and hammer peening [2–4].

Y. Wei and S. Chng (Eds.): INCASE 2021, LNME, pp. 88–92, 2022.
https://doi.org/10.1007/978-981-16-5763-4_19

The analysis of RS is essential to ensure the reliability of surface enhanced components under actual service conditions. Destructive RS characterization techniques such as hole-drilling produce material debris while a non-destructive X-Ray Diffraction (XRD) technique also becomes semi-destructive in nature when material layers are incrementally removed to characterize the sub-surface RS [5–7].

In this paper, a quantitative study of the Barkhausen noise (BN) technique was utilized to assess the effect of various shot peening process parameters on the BN signal measured. The correlation of BN profile and RS measured by XRD technique was investigated. Key BN features were discussed. This paper aimed to present the BN technique as a cost-effective, efficient as well as reliable non-destructive method for stress-depth profile in magnetic materials.

2 Materials and Methods

The material used for the study was Stainless Steel 440C (SS440C). Each specimen had a dimension of $7.5 \times 2 \times 1\ cm^3$ and were subjected to various shot-peening intensities of 0 mmA (as-received), 0.24 mmA, 0.312 mmA, 0.36 mmA and 0.45 mmA, using the conventional ASH 230 shots with 200% coverage at a mass flow rate of 3 kg/min.

BN measurements were conducted using a Rollscan 350 Barkhausen Noise Analyzer, with a measuring voltage and frequency of 5 V_{pp} and 125 Hz, respectively. A Microscan software was used for data collection, and computed various parameters from the Microscan BN burst graph as compiled in Table 1.

Table 1. Barkhausen noise (BN) parameters

Parameters [units]	Explanation
Root-mean-square (RMS) [mV]	Root mean square value of the BN signal
Peak value [mV]	Peak value of BN signal
Peak position [ms]	Peak position of BN signal
Full-width at half-maximum (FWHM) [ms]	Full Width at Half Maximum value of BN signal peak

Average values of RMS, FWHM, peak and pulse counts gathered from each specimen were analyzed and evaluated accordingly. To establish a correlation study between BN and RS, XRD (Stresstech, Xstress Robot) with assistance of electrochemical polishing (Struers) was utilized to get RS across the shot-peened surfaces at various depths. RS measurements were conducted at every incremental depth of 50 μm to establish a depth profiling of RS.

3 Experimental Results and Discussion

3.1 Barkhausen Noise Depth-Frequency Computation

The BN technique provides information from the sub-surface of the material. An estimation of the penetration depth (from sample surface) of BN signal is as shown in Eq. 1:

$$\delta = \frac{1}{\sqrt{\mu \pi \sigma f}} \tag{1}$$

Same as other noise signals, BN degrades with increasing of depth. Some work can be found to discuss the filter frequency and its influence in BN data collection. To our best knowledge, not much work has been done in generating depth-resolved BN and correlating with corresponding surface integrity properties [8, 9]. BN analysis depth was varied by altering analyzing frequencies, as shown in Table 2.

Table 2. Filter frequency to depth conversion

Depth [μm]	50	100	150	200	250	300	350	400
Filter frequency range [kHz]	500–1600	300–410	105–185	70–103	50–65	35–46	27–34	20–26

3.2 Barkhausen Noise Data Collection

Figure 1 (left vertical axis) shows the depth-resolved profiles of BN features (RMS, FWHM, peak and pulse counts) of the as-received and shot peened samples at various intensities. The highest values of RMS, peak and pulse counts are observed at shallower depth of 50 μm and the lowest are presented at deeper depth of 400 μm, as shown in Fig. 1a, c and d, respectively. For all the samples that were measured, the three aforementioned BN features decreased with increasing depth. This can be attributed to the presence of compressive residual stresses reduces along the increasing of depth, which is consistent with the fact that the shot-peening treatment gives significant impact in the near surface region than in its bulk.

For the FWHM (Fig. 1b), it shows a different trend from others, as the FWHM value first increases and then decreases at increasing depth. Compared with the others, the sample that was shot peened at 0.45 mmA has the highest value of RMS and peak, which generally, increases with shot peening intensity. While for the pulse count, limited difference can be seen among different samples.

The RS results are also included into Fig. 1 (right vertical axis) to facilitate its correlation analysis with BN features. An inversely proportional trend was observed in RS-RMS (Fig. 1a), RS-Peak (Fig. 1c), and RS-Pulse Counts (Fig. 1d) correlations, there were indistinguishable differences between the pulse counts registered at every

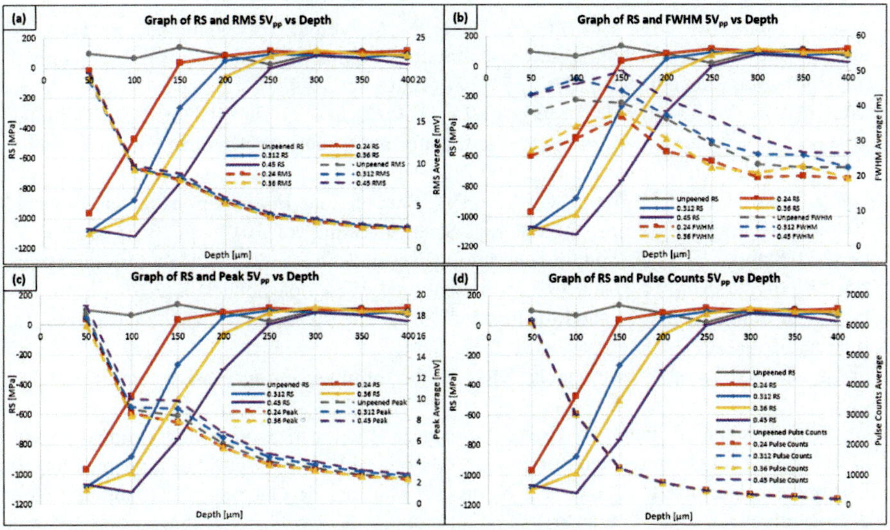

Fig. 1. Correlation graphs. (a) RS-RMS-depth, (b) RS-FWHM-depth, (c) RS-peak-depth, and (d) RS-pulse counts-depth.

incremental depth for each sample. RMS-depth and peak average-depth produced similar trends, while no clear correlation could be established in the case of RS-FWHM-depth (Fig. 1b). As a result, both RMS and peak show the best relationship when correlated to RS. From 50 μm to 250 μm, RMS values decreased as XRD RS value in the shot-peened SS440C increased, i.e. be less compressive. Beyond 250 μm where XRD RS values reached a plateau, RMS values were also observed to have stagnated.

4 Conclusion

In this paper, the correlation of BN key features (RMS, peak, FWHM and pulse counts) and RS was discussed. The RMS and peak were showing an inversely proportional trend comparing with the RS depth profile measured by XRD. These two features were deemed the best fits as significant trends of shot-peened SS440C with different intensities at various depths were illustrated, showing that BN is a promising technology for non-destructive residual stress measurement. Since the BN signal reveals a culmination of effects from surface integrity, the future study will be focused on the partitioning of the effects on BN signals thus enabling quantitative evaluation using BN technique.

Acknowledgement. This work is supported by an industry alignment fund-prepositioning programme: Advanced Post-processing and Non-Destructive Evaluation for Additive Manufacturing (IAF-PP, Grant no. A20F9a0045).

References

1. Sharpe, W.N., Jr.: Springer Handbook of Experimental Solid Mechanics. Springer, New York (2008). https://doi.org/10.1007/978-0-387-30877-7

2. Liu, H., Tan, C.K.I., Wei, Y., Lim, G., Cheng, W., Maharjan, N.: Robotic hammer peening-induced martensite in austenitic steels: spatial distributions of plastic deformation and phase transformation. Procedia CIRP **87C**, 172–177 (2020)
3. Liu, H., Wei, Y., Tan, C.K.I., Ardi, D.T., Tan, D.C.C., Lee, C.J.J.: XRD and EBSD studies of severe shot peening induced martensite transformation and grain refinements in austenitic stainless steel. Mater. Charact. **168**, 110574 (2020)
4. Lee, J.J., Wei, Y., Liu, H., Tan, C.C., Ardi, D.T.: Surface integrity variations of stainless steel 304 upon severe shot peening. In: Materials Science Forum, vol. 1015, pp. 30–35 (2020)
5. E837-20 Standard Test Method for Determining Residual Stresses by the Hole-Drilling Strain-Gage Method, ASTM E837-20. ASTM International, West Conshohocken (2020)
6. E2860-20 Standard Test Method for Residual Stress Measurement by X-Ray Diffraction for Bearing Steels. ASTM International (2020)
7. Fry, A.: Residual stress measurement: XRD depth profiling using successive material removal (2002)
8. Grum, J., Zerovnik, P., Fefer, D.: Use of Barkhausen effect in measurement of residual stresses in steel after heat treatment and grinding. In: Materials of 15th WCNDT Conference (2000)
9. Lasaosa, A., Gurruchaga, K., García Navas, V., Martínez-de-Guereñu, A.: Characterisation of in-depth stress state by magnetic Barkhausen noise on machined steel acquiring different frequency bands. In: Advanced Materials Research, vol. 996, pp. 373–379 (2014)

Application of Phase and Crystallographic Orientation Mapping for Materials Characterization

Siew Lang Teo, Dennis Cheng Cheh Tan, and Ming Lin(✉)

Institute of Materials Research and Engineering,
A*STAR (Agency for Science, Technology and Research), #08-03,
2 Fusionopolis Way, Innovis, Singapore 138634, Singapore
{sl-teo,dennis-tan,m-lin}@imre.a-star.edu.sg

Abstract. Materials fabrication and characterization are the foundations of materials research. Transmission Electron Microscope (TEM) is a very powerful and versatile instrument used for the characterization of materials at nanoscale. Compared to the Electron Backscattered Diffraction (EBSD), the Nano-Beam Diffraction (NBD) in TEM allows characterization of structural information with high spatial resolution at localized positions. Moreover, the precession electron beam can provide pseudo-kinematical diffractions and improve the accuracy for structural determination. Thus, using precession nano-beam diffraction technique, the phase of materials, crystallographic orientations, dislocations and lattice mismatches can be revealed at nanometer scale through scanning the electron beam at the area of interest.

Keywords: Nano-beam diffraction · Electron Backscattered Diffraction · Transmission Kikuchi diffraction

1 Introduction

The Electron Backscattered Diffraction (EBSD) and Transmission Kikuchi diffraction (TKD) are popular techniques to study the phase and crystallographic orientation mapping of crystalline materials. The measurements are usually carried out in a scanning electron microscope (SEM), and using Kikuchi patterns to identify the information required. The spatial resolution is around 10 nm [1].

In order to achieve a higher resolution mapping, precession nano-beam diffraction (P-NBD) in transmission electron microscope (TEM) has gained interests over the years as it provides solutions of crystal structures in nanoscale by acquiring electron diffraction patterns from rotating a tilted TEM electron beam at a cone-shaped angle at the area of interest [2]. Conventional dynamic diffraction significantly decreases the scattering intensity in TEM diffraction patterns, especially when the crystal is orientated at the zone axis. In the precession mode, the diffraction can be collected and shown in axis without tilting the crystal exactly along a particular zone axis. Using this technique, the dynamical effects are strongly reduced, and the intensity of reflections in the pattern can

Y. Wei and S. Chng (Eds.): INCASE 2021, LNME, pp. 93–96, 2022.
https://doi.org/10.1007/978-981-16-5763-4_20

follow the quasi-kinematical diffraction for the sample thickness up to 100 nm. With the combination of precession mode and NBD, both the phase and crystallographic mapping at nanoscale are obtained and these are complementary to EBSD and TKD characterizations. These collected diffraction patterns are then matched against the pre-calculated diffraction templates and analysed to reveal grain structures, orientation mapping, etc. [2–4]. Here, we introduce the applications and examples of P-NBD on materials analysis, and show the comparison between EBSD, TKD and NBD techniques.

2 Methods

Comparison of the different techniques was performed using a shot peened Ti64 coupon [5, 6]. EBSD analysis was achieved on the cross-section surface of the coupon, using a 20 kV electron accelerating voltage with a scan step size of 1 μm. Thin electron transparent slice from the top surface was prepared using Focus Ion Beam (FIB) for TKD and NBD analysis. Both techniques use scan step size of 10 nm, with 20 kV and 200 kV acceleration voltage respectively.

3 Results and Discussion

Figure 1 shows the analysis results of Ti64 coupon after a shot peening process. Shot peening is a cold working process in which the metal surface is hammered by metal shots and thus creating plastic deformation close to the surface. Figure 1a presents the cross-section SEM image of Ti64 after shot peening that is partially overlapped with a Kernal average misorientation (KAM) map obtained by EBSD. The orientation mapping from EBSD can be obtained for a large area. The bulk phase of Ti64 remains intact and large grains are seen. The grain size decreases from the bulk to the surface. The top surface after shot peening are shown in black, indicating that they are unindexed under EBSD analysis. This unindexed region could be due to the presence of amorphous structures, pores or voids, or small grains which are beyond EBSD resolution limit and thus unrecognized [5].

EBSD	TKD	NBD
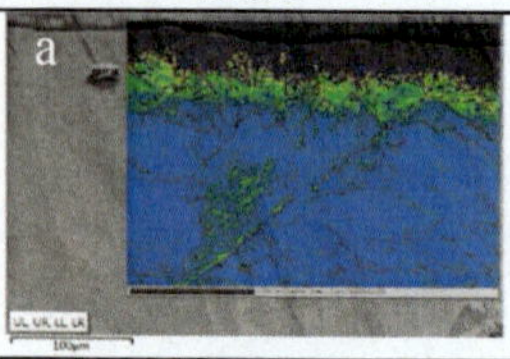	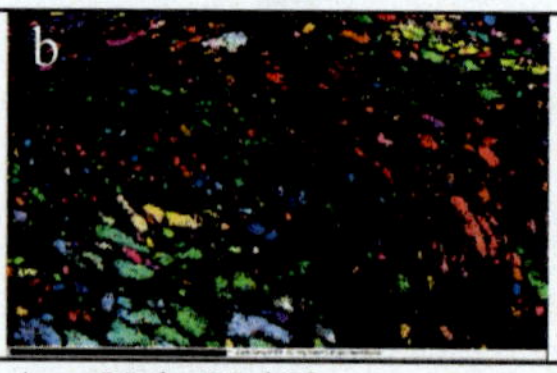	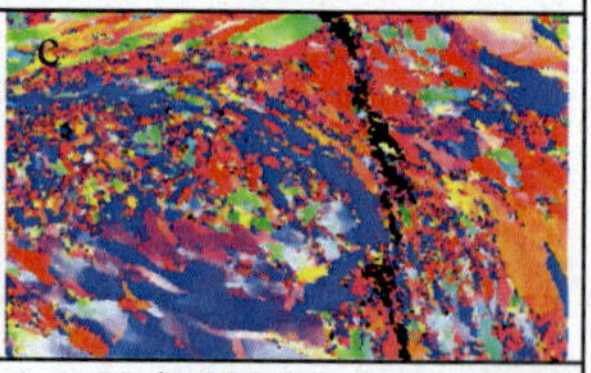
Acc Vol: 20 kV Scan area: 300×250 μm Step Size: 1 μm	Acc Vol: 20 kV Scan area: 5×3 μm Step Size: 10 nm	Acc Vol: 200 kV Scan area: 5×3 μm Step Size: 10 nm Precession Angle: 0.4°

Fig. 1. (a) KAM maps overlaid on the cross-sectioned SEM image (b & c) orientation map in z direction of Ti64.

To address these doubts, an electron transparent slice was cut and prepared by FIB for TKD and NBD analysis. Figure 1b and 1c were taken at the same area of the sample. Although a few grains structures can be identified by TKD, most of the area is still dark and unidentified, probably owing to the poor quality Kikuchi patterns from lower acceleration voltage and larger electron beam size in SEM. However, NBD mapping using TEM is able to identify all areas unambiguously. From TEM phase and crystallographic mapping, the α-Ti grains and β-Ti strips (in black, crystal orientation is not shown here) can be indexed clearly. It is shown that the original large grains had been severely deformed and had broken into smaller ones with sizes from a few tens to a few hundreds of nanometres after shot peening had been applied on the surface.

From the example shown above, we can summarize the comparison of EBSD, TKD and NBD techniques in Table 1. SEM-EBSD can obtain information on large specimen but requires a flat and highly polished surface due to electron diffraction from backscattered electrons. The resolution can reach 20 nm using field emission SEM, but is difficult to be achieved on most samples due to poor quality Kikuchi patterns from small grains. TKD and NBD use transmitted elections, thus electron transparent samples are required which limits the scan area of the specimen. The resolution is much higher than EBSD and the grains as small as 10 nm can be easily shown by NBD mapping.

Table 1. Comparison of EBSD, TKD and NBD techniques.

	EBSD-SEM	TKD-SEM	NBD-TEM
Data collection method	Kikuchi patterns	Kikuchi patterns	Electron diffraction patterns
Sample type	Bulk sample with smooth and flat surface	<100 nm thin specimen	<100 nm thin specimen
Scan area	μm to mm	nm to μm	nm to μm
Spatial resolution	20 nm	10 nm	1 nm

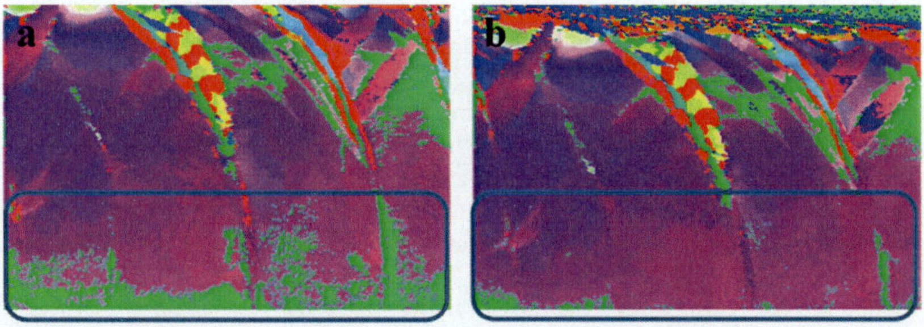

Fig. 2. Crystallographic orientation map in z direction of Ti64 (a) without precession and (b) with precession angle of 0.4°

Figure 2 shows the TEM crystallographic orientation map of Ti64 sample with and without beam precession. Some noise is found at the bottom of the map without precession mode, which is highlighted by the rectangle. Typically, the bottom part of FIB prepared sample is thicker than the top, thus less electrons can pass through and less/weaker diffraction spots are produced. Hence, the diffraction is difficult to be identified due to limited presence of diffraction spots in the pattern. On the contrary, the precession mode can accumulate all the electron patterns when the beam is scanned by rotating in a cone-shape. More diffraction spots are obtained and the improved quality of the pattern can make the identification more accurate. In this case, the α-Ti grains can be identified and noise is minimized with precession mode.

4 Conclusion

Precession nano-beam electron diffraction in TEM plays a substantial role for full analysis of nanostructured materials, including high quality nanoscale phase and orientation mapping and achieves better resolution for smaller grain samples. It is complementary to SEM-EBSD and a combination of both would provide a comprehensive characterization of materials structures from bulk to nanometer scale.

Acknowledgement. This research is supported by the Agency for Science, Technology and Research (A*STAR) under the Structural Metal Alloys Programme (A18b1B0061).

References

1. Oxford Instruments. http://www.ebsd.com/ebsd-applications/ebsd-applications-by-technique/tkd. Accessed 10 June 2021
2. Brons, J.G., Thompson, G.B.: Orientation mapping via precession-enhanced electron diffraction and its applications in materials science. JOM **66**(1), 165–170 (2013). https://doi.org/10.1007/s11837-013-0799-5
3. Darbal, A.D., Gemmi, M., Portillo, J., Rauch, E., Nicolopoulos, S.: Nanoscale automated phase and orientation mapping in the TEM. Microsc. Today **20**(6), 38–42 (2012)
4. Rauch, E.F., Véron, M.: Automated crystal orientation and phase mapping in TEM. Mater. Charact. **98**, 1–9 (2014)
5. Zhang, Z., et al.: Fatigue life enhancement in alpha/beta Ti-6Al-4V after shot peening: an EBSD and TEM crystallographic orientation mapping study of surface layer. Materialia **12**, 100813 (2020)
6. Lim, S.H., et al.: Warm shot peening on Ti-6Al-4V alloy: effects of temperature on fatigue life, residual stress, microstructure and mechanical properties. J. Alloy. Compd. **882**, 160701 (2021)

Damage Assessment of Helicoidal CFRP Using Active Infrared Thermography

Jonathan Zheng(✉), Carlos Manzano, Vinod Kumar, and Andrew Ngo

Institute of Materials Research and Engineering, 2 Fusionopolis Way, Innovis, #08-03, Singapore 138634, Singapore
{jonathan_zheng,garciac,vinod_kumar,ngocya}@imre.a-star.edu.sg

Abstract. Composites have been gaining popularity in recent years, being utilized not only in the automobile industry but in aerospace and the military as well. With an increased utilization of composites comes the need for cost-effective and efficient Non-Destructive Testing (NDT) techniques to assess their quality after manufacturing and especially during maintenance processes. Active infrared thermography comes in with the ability to provide a fast and reliable inspection technique for composites while maintaining cost effectiveness. This study looks at the use of active thermography techniques (lock-in, transient, flash) to assess the damages sustained by helicoidal structured CFRP. The helicoidal CFRP samples undergo active thermography tests after being subjected to a 35J drop-weight impact test (Reference: ASTM D7136). The results from the active thermography tests are then compared and discussed, highlighting flash thermography to be the most suitable active thermography method to use for the helicoidal CFRP samples in this study.

Keywords: CFRP · Drop-weight impact · Active-thermography · NDT

1 Introduction

The use of composites has been gaining an increasing popularity in recent years, being utilized not only in the automobile industry, but in aerospace and the military as well. With an increase in the utilization of composites in industries, comes the need for more cost effective and efficient NDT techniques to assess their quality after manufacture and especially during maintenance processes. Traditional NDT techniques such as ultrasonic provide a qualitative assessment of the composite at the expense of high costs due to the long man hours required for point by point inspections [1], while others like eddy current testing are limited to electrically conductive materials [2] and magnetic particle testing is limited to ferromagnetic materials [3]. The use of active infrared thermography for defect detection in composites has gained more popularity over the years as it is an efficient, non-contact, reliable and cost-effective method [4]. Helicoidal structured composites are bio-inspired composites that are made through a series of successive fibre layers with angle rotations between each adjacent ply. In static loading conditions, helicoidal composites have exhibited high stiffness, strength and toughness properties and have

Y. Wei and S. Chng (Eds.): INCASE 2021, LNME, pp. 97–100, 2022.
https://doi.org/10.1007/978-981-16-5763-4_21

proven to have increased impact absorption capabilities in dynamic tests [5–7]. In this experiment, the helicoidal CFRPs have gone through a 35J drop-weight impact test (using ASTM-D7136 as reference) and will have their damage assessed and compared through the use of the different active thermography techniques; lock-in, transient, and flash thermography.

2 Methods

36-Ply helicoidal composites of different stacking orientations; $[0, 22.5, \ldots, 180]_{2s}$ and $[0, \ldots, 180]_{3s}$ (22.5° & 36°), were fabricated through the hand layup method using plain weave carbon fibre. They were subjected to a 35J drop-weight impact test based on ASTM-D7136 as a guiding reference (Fig. 1).

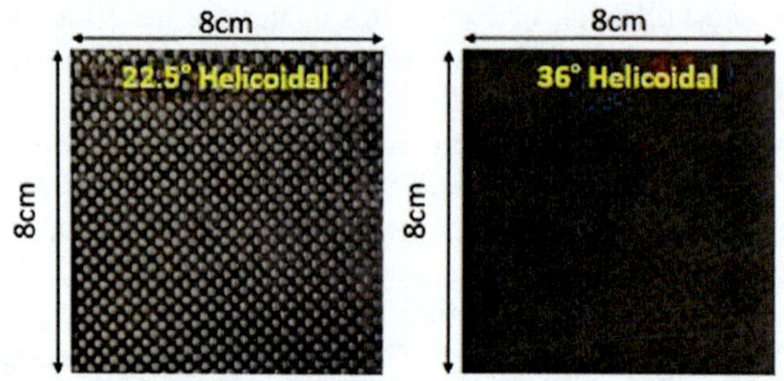

Fig. 1. Helicoidal carbon fibre composite samples 22.5° (left) and 36° (right) samples of 8 cm by 8 cm dimensions with nominal thickness of 6.3 mm

Transient, lock-in and flash thermography methods were utilized to assess the damages sustained by the helicoidal composites. Different parameters were tested for each method to ensure the best thermal images were obtained (Fig. 2 and Table 1).

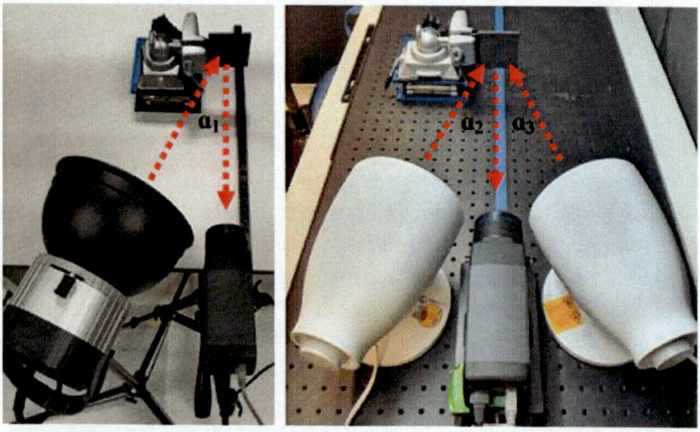

Fig. 2. Experimental setup of flash (left) and transient/lock-in thermography (right) where $\alpha_1 = 18°$ and $\alpha_2 = \alpha_3 = 15°$

Table 1. Selected experiment parameters

Selected experiment parameters			
Method	Excitation	Duration	Distance
Transient	2 × 150W IR	30 s Heating & Cooling, Total: 60 s	0.4 m
Lock-In	2 × 150W IR	3 Cycles, 40 s/Cycle, Total: 120 s	0.4 m
Flash	1 × 6kJ Flash	Single 2 ms Flash, 15 s Acquisition, Total: 15 s	0.4 m

3 Results and Discussion

See Table 2.

Table 2. Active thermography inspection results of transient, lock-in and flash

Active Thermography Results				
	Visual	Transient	Lock-In	Flash
22.5° CFRP	 Dimension: 8x8 cm Thickness: 6.3mm	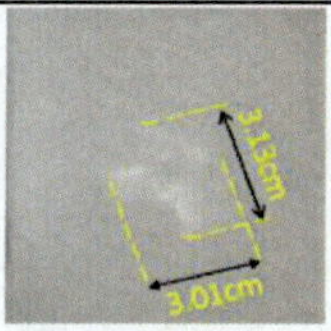 Defect Length: 3.13cm Defect Width: 3.01cm	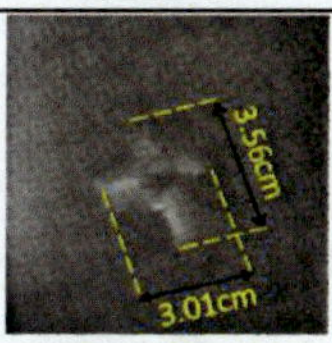 Defect Length: 3.56cm Defect Width: 3.01cm	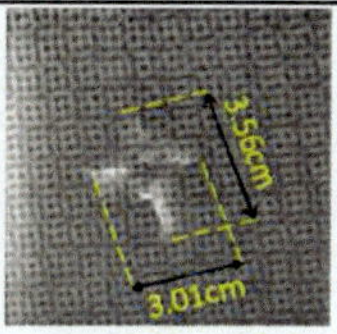 Defect Length: 3.56cm Defect Width: 3.01cm
36° CFRP	 Dimension: 8x8 cm Thickness: 6.3mm	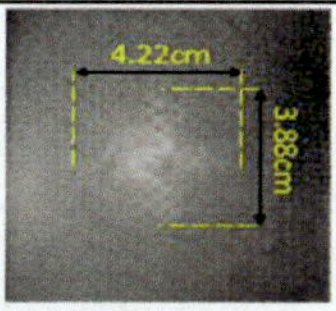 Defect Length: 3.88cm Defect Width: 4.22cm	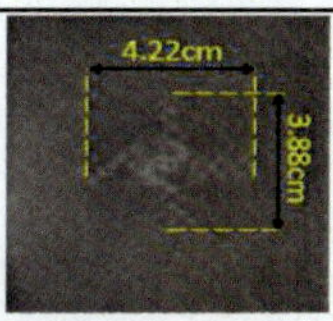 Defect Length: 3.88cm Defect Width: 4.22cm	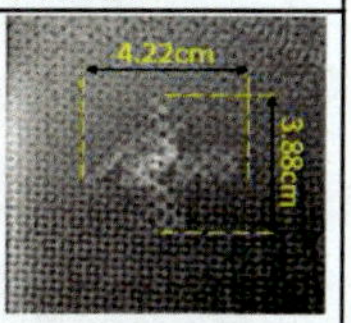 Defect Length: 3.88cm Defect Width: 4.22cm

From the results obtained, the active thermography methods were able to reveal subsurface defects caused by the drop-weight impact which were not visible to the naked eye. Based on the length and width of the defect, the 36° helicoidal CFRP seems to have sustained greater damage than the 22.5° helicoidal CFRP. This is however not conclusive to determine which helicoidal CFRP sustained a greater damage as it only shows damage in the lateral plane without taking into account the actual depth of the damage. Lock-in and flash thermography were able to show the extent of damage in both the length and width wise while transient thermography failed to reveal the full extent of the subsurface defect length wise in the 22.5° helicoidal composite. This might have been caused by reaching thermal equilibrium on the surface layers of the sample during transient heating, making transient thermography limited based on the sample's thermal diffusivity. Flash thermography provided the best defect contrast as well as the fastest in execution time as compared to the other methods. It is however limited by its penetration depth into the CFRP sample as compared to lock-in thermography whereby the parameters may be adjusted for even deeper defect depth detection.

4 Conclusion

Transient, lock-in, and flash thermography were used to assess the damages sustained by the helicoidal CFRPs. Based on the length and width of the defects detected, the 36° helicoidal CFRP appears to have sustained a greater amount of damage than the 22.5° helicoidal CFRP but this is not conclusive as the real extent of the depth of damage was not truly determined. Flash thermography provided the fastest and best defect contrast compared to transient and lock-in thermography making it the most efficient and would be the most suitable method to employ in the inspection of these 6.3 mm thick plain weave helicoidal CFRPs.

Acknowledgements. This research work was supported by the Agency for Science, Technology and Research (A*STAR) through grants under its Singapore Aerospace Programme Cycle 13 (Grant No. A1915a0072) and Polymer Matrix Composites Programme (Grant No. A19C9a0044). Support of Singapore's National Research Foundation and Building & Construction Authority is also acknowledged (Grant No. NRF2018NRF-CG002-005).

References

1. Nondestructive Evaluation Techniques: Ultrasound. https://www.nde-ed.org/NDETechniques/Ultrasonics/ultrasound.xhtml. Accessed 23 Apr 2021
2. Introduction to Eddy Current Testing: Olympus IMS. https://www.olympus-ims.com/en/eddycurrenttesting/. Accessed 21 Apr 2021
3. Ibrahim, M.E.: Non-destructive inspection and structural health monitoring of aerospace materials. Maritime Division, Defence Science and Technology Organisation, Melbourne (2016)
4. Sarawade, A.A., Charniya, N.N.: Infrared thermography and its applications: a review. In: 3rd International Conference on Communication and Electronics Systems (ICCES), pp. 280–285 (2018). https://doi.org/10.1109/CESYS.2018.8723875
5. Grunenfelder, L.K., et al.: Bio-inspired impact-resistant composites. Acta Biomater. **10**(9), 3997–4008 (2014). https://doi.org/10.1016/j.actbio.2014.03.022
6. Andersons, M.K.J.: Dependence of fracture toughness of composite laminates on interface ply orientations and delamination growth direction. Compos. Sci. Technol. **64**, 2139–2152 (2004)
7. Shang, J.S., et al.: Crustacean-inspired helicoidal laminates. Compos. Sci. Technol. **128**, 222–232 (2016). https://doi.org/10.1016/j.compscitech.2016.04.007

In Situ High Temperature XRD Studies of Molten Natural CMAS Corrosion on Thermal Barrier Coatings

S. H. Lim[1(✉)], C. S. Chua[1], Brian T. M. Ong[1,2], Coryl Jing Jun Lee[1], Dennis Cheng Cheh Tan[1], L. T. Koh[2], W. K. Na[2], and Shijie Wang[1]

[1] Institute of Materials Research and Engineering (IMRE), A*STAR (Agency for Science, Technology and Research), 2 Fusionopolis Way, Singapore 138634, Singapore
{limsh,chuacs,coryl-lee,dennis-tan,sj-wang}@imre.a-star.edu.sg, Thiamminbiran.ong@ge.com

[2] ST Engineering Aerospace Engines Pte Ltd., 501 Airport Rd, Singapore 539931, Singapore
{liteck,wkna}@stengg.com

Abstract. High temperature behaviors of natural calcium-magnesium-alumino-silicate (CMAS) and CMAS coated thermal barrier coating (TBC) systems (CMAS-TBC) were investigated using combined techniques of in-situ high temperature XRD-Rietveld quantitative phase analysis and elemental mapping analysis. The XRD result indicates that silicate mineral, diopside ($CaMgSi_2O_6$), is formed at temperature higher than 800 °C. At 1200 °C, CMAS is found to have penetrated into the YSZ layer resulting in chemical interaction between CMAS and YSZ. Prolonged heat treatment shows that the interaction results in the formation of zircon ($ZrSiO_4$) and spinel ($MgAl_2O_4$). This thermochemical interaction results in the weakening of the YSZ layer, which ultimately leads to the spallation of YSZ layer during the cooling down phase of the thermal cycling.

Keywords: CMAS · Corrosion · Thermal barrier coating · In situ XRD

1 Introduction

The application of thermal barrier coating (TBC) on aircraft engine components has been critical in increasing the lifespan of the components, preserving structural health integrity of our aircraft engines, and achieving better fuel efficiency for our air flight [1, 2]. The lifespan of TBC layer has, however, been limited by the ingestion of calcium-magnesium-alumino-silicate (CMAS) particles. Such ingestion is especially prominent at elevated temperatures where melting and penetration of CMAS into the TBC layer is occurring [3–5]. In this article, high temperature behavior of natural silicate dust (CMAS) attack on TBC system, combined with in-situ XRD technique, were investigated to understand the mechanism of molten CMAS infiltration into TBC under operating conditions.

Y. Wei and S. Chng (Eds.): INCASE 2021, LNME, pp. 101–104, 2022.
https://doi.org/10.1007/978-981-16-5763-4_22

2 Material and Methods

The base substrate, Inconel 718, was coated with a bond coat layer, NiCrAlY (~200 μm thickness, Amdry 962, Oerlikon Metco) and ceramic top layer, 7% YSZ powder (~300–500 μm thickness, METCO 204NS) using atmospheric plasma spray coating (APS, Sulzer Metco Plasma Spray System). Natural CMAS powder, scraped from turbine blades, was mixed in a terpineol-based organic solvent (Fuel Cell Materials) at a weight ratio of 1 part solvent to 1.75 part CMAS powder. The mixture was coated onto the TBC coupon using a spatula, and dried overnight.

The morphology and elemental mapping analysis of natural CMAS powder and cross-section of CMAS-TBC coupons were carried out using Energy-dispersive X-ray spectroscopy – FESEM (Joel JSM 7600F). In situ phase evolution of CMAS-TBC coupon was monitored using X-ray Diffraction (XRD, Bruker D8 Advance, Cu Kα source) with a high temperature chamber (Anton Parr HTK1200) operated at 40 kV and 40 mA. All quantitative phase analysis was performed using Rietveld method as implemented in TOPAS V5. The coupon was heated from room temperature to 1200 °C at the heating rate of 5 °C per minute, and followed by 5 cycles of Isothermal Oxidation Test (IOT) with each cycle at 1200 °C for 5 h.

3 Results and Discussion

3.1 Phase Evolution and Microstructure of CMAS-TBC

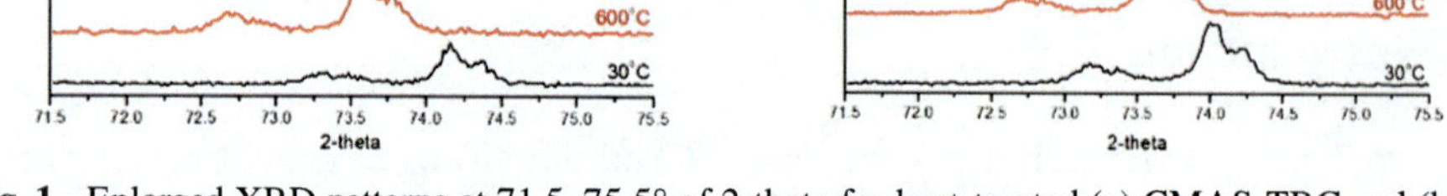

Fig. 1. Enlarged XRD patterns at 71.5–75.5° of 2-theta for heat-treated (a) CMAS-TBC and (b) TBC coupons. t-YSZ: tetragonal YSZ.

Comparing the enlarged XRD patterns (Fig. 1), peak shift of t-YSZ (2 2 0) and (0 0 4) for pristine TBC and CMAS-TBC coupons is similar that the shift to lower 2-theta angle is observed due to temperature-dependent lattice expansion. At 1200 °C, the reflection peak of tYSZ (0 0 4) for CMAS-TBC coupon is weakened and broadened. Meanwhile, diffuse scattering around the peaks of t-YSZ (2 2 0) and (0 0 4) is noticed as well, which is attributed to non-uniform lattice distortion of t-YSZ. This phenomenon is more significant after cooling down to room temperature, and indicates that molten CMAS penetrates into the porous structure of t-YSZ at 1200 °C.

3.2 Phase Evolution and Microstructure of CMAS-TBC

In situ XRD analysis of IOT heating reveals the formation of magnesium aluminate spinel ($MgAl_2O_4$), corundum (α-Al_2O_3) and zircon ($ZrSiO_4$). Table 1 shows that with each successive IOT cycle, there is a general decrease in Rietveld wt % for t-YSZ and an increase for $MgAl_2O_4$, $ZrSiO_4$, and α-Al_2O_3. The increase of α-Al_2O_3 wt % indicates the growth of thermal grown oxidation (TGO) layer (Fig. 2) at the interfacial top coat (TC)/bond coat (BC) region [3] after each IOT heating.

Table 1. Phase composition of CMAS-TBC coupon after 5 cycles (C) of IOT. t-YSZ: tetragonal YSZ; m-ZrO_2: monoclinic ZrO_2.

Heating	Rietveld calculated weight fraction (wt %)								
	TBC		CMAS				Others		
	t-YSZ	m-ZrO_2	Silicate minerals	Fe_2O_3 Fe_3O_4	$CaCO_3$	$CaSO_4$	$MgAl_2O_4$	α-Al_2O_3	$ZrSiO_4$
30 °C	60.64	1.61	27.48	1.04 (Fe_3O_4)	0.78	8.45	–	–	–
800 °C	56.22	1.78	33.80	3.79	–	4.43	–	–	–
1200 °C	91.06	1.20	3.14	3.60	–	–	–	–	–
C1	77.85	1.49	4.78	3.33	–	–	0.16	12.14	0.25
C3	63.41	0.90	6.50	2.58	–	–	2.46	22.01	2.14
C5	52.81	0.86	9.45	2.18	–	–	4.43	27.60	2.66

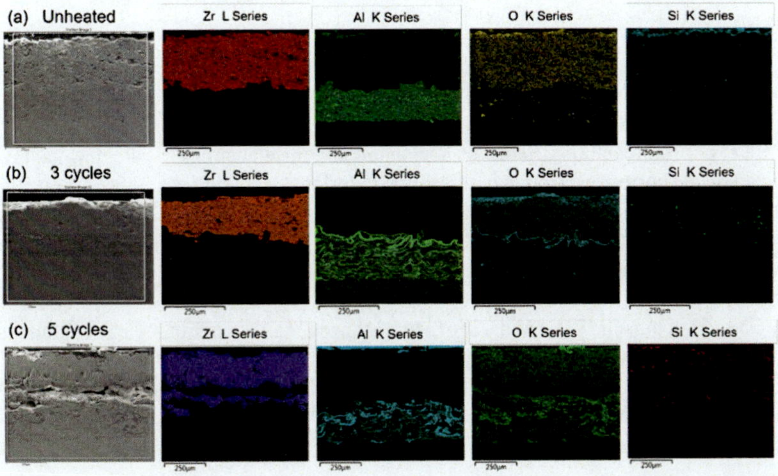

Fig. 2. EDX elemental mapping analysis of (a) unheated, (b) 3 cycles and (c) 5 cycles of IOT CMAS-TBC coupons that shows the growth of TGO in the interfacial region of TC/BC.

The formation of $ZrSiO_4$ after first cycle of IOT evidences that the diffusion of Si from amorphous molten CMAS, in which less crystallite silicate minerals can be detected using XRD analysis, starts after prolonged 5 h heating at 1200 °C. For $MgAl_2O_4$, it is formed in the BC layer through the diffusion of Mg element from CMAS. The formation of these species is accompanied with the spallation of TC layer resulting in the decrease in t-YSZ phase. The increase in α-Al_2O_3 or TGO layer occurs only after cooling down of the CMAS-TBC coupon from C1 to C5. This indicates that further spallation occurs during the cooling down phase, hence, exposing more underlying TGO layer for XRD detection.

4 Conclusion

In this work, in situ high temperature XRD is used to probe the thermochemical reaction between natural CMAS and TBC during high temperature operating conditions. Our XRD and STA findings show that several silicate minerals, such as quartz, albeit high and anorthite, were detected before the appearance of diopside $CaMgSi_2O_6$ at 800 °C. With the penetration of molten CMAS at 1200 °C, further thermochemical interaction took place between diopside and TBC layer, forming $ZrSiO_4$ and $MgAl_2O_4$ phase. This further weakens the TBC structure, which ultimately leads to its spallation and failure during the cooling down stage.

Acknowledgements. This research is supported by the Agency for Science, Technology and Research (A*STAR) under the Accelerated Materials Development for Manufacturing (A1898b0043), A*STAR Singapore Aerospace Programme (AP) and ST Aerospace Engineering Pte Ltd.

References

1. Clarke, D.R., Oechsner, M., Padture, N.P.: Thermal-barrier coatings for more efficient gas-turbine engines. MRS Bull. **37**(10), 891–898 (2012)
2. Zhu, C., et al.: Improved re-sistance of lanthanum zirconate coatings to calcium-magnesium-alumina-silicate corrosion through composition tailoring. Ceram. Int. **44**, 13908–13915 (2018)
3. Nieto, A., Agrawal, R., Bravo, L., Hofmeister-Mock, C., Pepi, M., Ghoshal, A.: Calcia–magnesia–alumina–silicate (CMAS) attack mechanisms and roadmap towards Sandphobic thermal and environmental barrier coatings. Int. Mater. Rev. https://doi.org/10.1080/09506608.2020.1824414
4. Levi, C.G., Hutchinson, J.W., Vida-Sétif, M., Johnson, C.A.: Environmental degradation of thermal barrier coatings by molten deposits. MRS Bull. **37**, 932–941 (2012)
5. Li, L., Hitchman, N.J., Knapp, K.: Failure of thermal barrier coatings subjected to CMAS attack. J. Therm. Spray Technol. **19**(1), 148–155 (2010). https://doi.org/10.1007/s11666-009-9356-8

Effect of Build Angle on Surface Roughness of High-Speed Selective Laser Melted Ti-6Al-4V Alloy

Eunhyeok Seo[1(✉)], Namhun Kim[1], Seung Ki Moon[2], Hyokyung Sung[3], and Im Doo Jung[1(✉)]

[1] Ulsan National Institute of Science and Technology, UNIST-Gil 50, Ulsan 44919, Republic of Korea
{okeh1216,nhkim,idjung}@unist.ac.kr

[2] Nanyang Technological University, 50 Nanyang Ave, Singapore 639798, Singapore
skmoon@ntu.edu.sg

[3] Gyeongsang National University, Jinju 52828, Republic of Korea
hksung@gnu.ac.kr

Abstract. Despite recent progress in achieving high mechanical properties of metal additive manufacturing, the low productivity represented by slow manufacturing speeds still remains a bottleneck of the technology. We improved the scanning speed of selective laser melting (SLM) with Ti-6Al-4V up to 1800 mm/s for high printing speed with affordable mechanical properties. In addition, the hot isostatic pressing (HIP) process was applied as post-treatment process to remove porosity of the high-speed printed SLM specimens. However, the HIP process caused a microstructural change from α′-lath martensite to a Widmanstätten α-lamellar structure. This microstructural change was most pronounced on the surface area and caused inter-lamellar fracture. Here, we investigate this phenomenon with microstructural and computational analysis. The results indicate that this deterioration phenomenon is highly dependent on surface roughness, which can be controlled with a build angle set-up.

Keywords: Selective Laser Melting · Ti-6Al-4V · Scanning speed · Laser power · Tensile properties · Surface roughness

1 Introduction

Additive manufacturing (AM) with metals has been actively studied in many industrial fields [1, 2]. The slow manufacturing speed of the metal AM process is the biggest obstacle to its development [3]. If the metal AM process significantly enhances its manufacturing speed, it can cause the global manufacturing industry to move beyond conventional manufacturing methods.

In this regard, there have been active studies on high-speed additive manufacturing processes [4]. SLM is one of the AM techniques that fabricates 3D shapes with melting the metallic powders via laser beam. Recent studies report that the mechanical strength

Y. Wei and S. Chng (Eds.): INCASE 2021, LNME, pp. 105–108, 2022.
https://doi.org/10.1007/978-981-16-5763-4_23

of SLM has reached similar level with conventional metal forming methods, but its productivity is still relatively low [5]. One of the ways to increase productivity is to speed up scanning. If the scanning speed is high in SLM, the unmelted powder will inevitably increase, and the mechanical properties will decrease due to many pores. To avoid this phenomenon, a high enough energy density laser power must be accompanied. Even with high laser power, micropore formation is hard to avoid due to volume shrinkage during the SLM process from liquid to solid. To reduce these micropores, the hot isostatic pressing (HIP) process is essential [6, 7].

In this paper, the HIP process effectively reduced the micropores, but resulted in poor mechanical properties. The reason for this reverse effect is that micro-tissue changes have affected the surface roughness. As an alternative to reduce the reverse effect, we investigated the effect of improving the mechanical properties according to the build angle setting.

2 Methods

Ti-6Al-4V powders were used for the SLM process, and the specimens were manufactured with a variety of process parameters (laser power: 245–343 W, scanning speed: 1250–1750 mm/s, hatch space: 70 μm, and layer thickness: 30 μm) using Pro X320 DMP. 'H' specimen means that supporter height is 30 mm, and the 'L' specimen means that supporter height is 10 mm from the plate. The HIP process was performed at 950 °C for 3 hours under a 100 MPa pressure in a 99.999% Ar environment, which was followed by furnace cooling. The tensile specimens were manufactured with a total length of 75 mm, a diameter of 4.75 mm, and a gauge length of 25.4 mm. These specimens were built with built angles of 0°, 45°, and 90°.

3 Results and Discussion

Ti-6Al-4V tensile specimens were printed with high scanning speed and laser power. The printed specimens were divided into as-built specimens and hip'd specimens. In the x-ray micro-computed tomographic analysis, porosity tends to be very low, 0.02% to 0.04% (Fig. 1a). Unlike As-built specimens, which has no post-processing, Hip'd specimens have significantly reduced its pore. However, after the Hip process, the mechanical properties decreased, especially the elongation decreased significantly. The reason for this reverse effect was due to changes in microstructure. The microstructure changed from α'-martensite to Widmanstätten after HIP process.

To reduce this reverse effect, we set the build angle as a way to control surface roughness. We fabricated the three specimens with build angles of 0°, 45°, 90°, and performed the HIP process. Table 1 shows tensile test results for the three specimens. In all the specimens, there was no significant difference in yield strength and tensile strength. However, there is a big difference between the three specimens' elongation. In a 90° specimen, the elongation was 5%, while other specimens were less than 2%.

Three structures with different surface roughness were modeled. Figure 2 shows the rectangular-type structure of three different surface roughness configurations: smooth, medium-roughness, and high-roughness specimens. A tensile force of 100 kN was

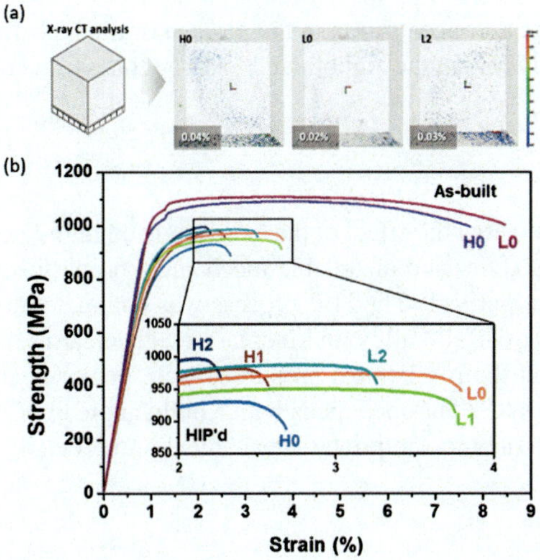

Fig. 1. (a) X-ray CT analysis for the inner micro-porosity measured for the H0, L0, and L2 specimens. (b) Tensile graph for the as-built and the specimens subjected to HIP [8].

Table 1. Tensile properties of the SLM Ti-6Al-4V specimens with a different build orientation.

Specimen	Yield strength (MPa)	Tensile strength (MPa)	Elongation (%)
H0	798	930	1.7
H45	825	957	1.0
H90	811	1011	5.0

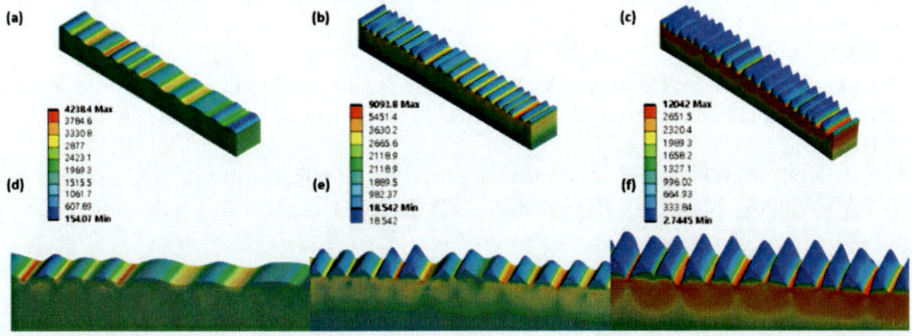

Fig. 2. Von-Mises stress distribution for (a)–(c) the overall tendency, and (d)–(f) the top surface stress distribution for three specimens [8].

applied on the wide side surface for each structure, and a fixed boundary condition was given on the narrow side for the mechanical analysis. The results of simulation indicated that the ridge area for each specimen has a relatively low von Mises stress, and there is a clear stress concentration in the pit area of the high-roughness specimen; however, it is relatively low in the smooth and low-roughness specimens.

4 Conclusion

In this study, we investigated the effect of the build angle on high-speed SLM Ti-6Al-4V to achieve high productivity and affordable mechanical properties. The results can be summarized into three parts: 1. The HIP process was conducted to effectively reduce micropores of SLM Ti-6Al-4V alloy and increase mechanical properties. 2. However, the results showed that the mechanical properties were poor, which is directly related to the surface roughness. 3. The specimens with a build angle of 90° showed the lowest surface roughness and the most improved mechanical properties.

References

1. Chen, Y., et al.: Usefulness of three-dimensional printing of superior mesenteric vessels in right hemicolon cancer surgery. Sci. Rep. **10**(1) (2020). Article number: 11660
2. Abdullah, F., Okuyama, K., Morimitsu, A., Yamagata, N.: Effects of thermal cycle and ultraviolet radiation on 3D printed carbon fiber/polyether ether Ketone Ablator. Aerospace **7**(7), 95–123 (2020)
3. Nazir, A., Jeng, J.Y.: A high-speed additive manufacturing approach for achieving high printing speed and accuracy. Proc. Inst. Mech. Eng. C **234**(14), 2741–2749 (2020)
4. Zavala-Arredondo, M., et al.: Laser diode area melting for high speed additive manufacturing of metallic components. Mater. Des. **117**, 305–315 (2017)
5. Jamshidi, P., et al.: Selective laser melting of Ti-6Al-4V: the impact of post-processing on the tensile, fatigue and biological properties for medical implant applications. Materials **13**(12), 2813–2829 (2020)
6. Bocanegra-Bernal, M.H.: Hot isostatic pressing (HIP) technology and its applications to metals and ceramics. J. Mater. Sci. **39**(21), 6399–6420 (2004). https://doi.org/10.1023/B:JMSC.0000044878.11441.90
7. Persenot, T., Martin, G., Dendievel, R., Buffiére, J.Y., Maire, E.: Enhancing the tensile properties of EBM as-built thin parts: effect of HIP and chemical etching. Mater. Charact. **143**, 82–93 (2018)
8. Lee, J., et al.: Reverse effect of hot isostatic pressing on high-speed selective laser melted Ti–6Al–4V alloy. Mater. Sci. Eng. A. **807**, 140880 (2021)

A Method of Dynamic Active Infrared Thermography to Reduce Inspection Time of CFRP Components

Vinod Kumar(✉), Carlos Manzano, Jonathan Zheng, and Andrew Ngo

Institute of Materials Research and Engineering, 2 Fusionopolis Way, Innovis, #08-03, Singapore 138634, Singapore
{vinod_kumar,garciac,jonathan_zheng,ngocya}@imre.a-star.edu.sg

Abstract. The accelerated use of carbon fibre reinforced plastics (CFRP) warrants the need for faster and lower-cost inspection techniques to verify the quality of manufactured products. For this purpose, non-destructive testing (NDT) techniques are commonly used and active infrared thermography (IRT) is a highly recommended method. In most applications, active IRT is performed in a static configuration where the elements involved in the inspection do not move. This presents some shortcomings when inspecting larger surfaces or multiple identical components. An alternative approach is a dynamic inspection, which allows for shorter inspection cycle times when inspecting multiple identical components. This work analyses the thermographic results from the static IRT configuration and determines a methodology to achieve a suitable dynamic IRT configuration. The results demonstrate the feasibility of the proposed alternative approach in reducing inspection cycle times when inspecting multiple identical components.

Keywords: Non-destructive testing · Active thermography · Infrared thermography · NDT · CFRP

1 Introduction

Carbon fibre reinforced plastics (CFRP) are prominently used in several industries such as aviation, automotive, building construction and maritime transport [1].With the accelerated use of such fibre reinforced composite materials, these is a definitive need for effective and cost-efficient inspection techniques to ensure the high quality standards of manufactured products and for manufacturers to reach a competitive advantage.

Non-destructive testing (NDT) methods such as active infrared thermography (IRT) are used in a wide variety of applications including the inspection of CFRP components [2]. Active IRT is where an external excitation source is used to generate a thermal flow in the object of interest; regions with subsurface anomalies produce thermal contrast, which is reflected in the temperature-time response to the thermal stimulation and can be detected via an infrared (IR) camera [3].

Active IRT is often performed in a static configuration (SC) where all elements, *i.e.*, the IR camera, excitation source and object of interest do not move with respect to each

Y. Wei and S. Chng (Eds.): INCASE 2021, LNME, pp. 109–113, 2022.
https://doi.org/10.1007/978-981-16-5763-4_24

other [4]. This configuration of inspection presents some challenges when inspecting larger surfaces as it requires multiple inspections to be done and merging the results obtained. The static active IRT is also particularly time consuming when a large number of identical components are required to be inspected as each component would have to remain still for the entire duration of the inspection before moving on to the next component. To reduce inspection time in such scenarios, a possible alternative approach is a method of dynamic inspection where the sample is on a conveyor belt and moves from the excitation source to the IR camera.

2 Methodology

An active IRT inspection in a SC is performed on a CFRP sample to derive a set of characteristic cooling curve data which serves as the baseline. The baseline data is analysed and used to determine the setup for the active IRT dynamic configuration (DC). After which the CFRP sample is inspected using the DC and the data obtained is compared with that of the baseline data. The resulting comparison is evaluated to determine the feasibility of the DC for active IRT inspection.

2.1 Material Characterization

The specimen is a CFRP panel measuring 144 × 120 mm in size and 5.5 mm thickness with six perforations each of ø30 mm at varying depths as seen in Fig. 1. The perforations act as simulated defects and for the purposes of this experiment, the focus will be on Defects no. 1–4.

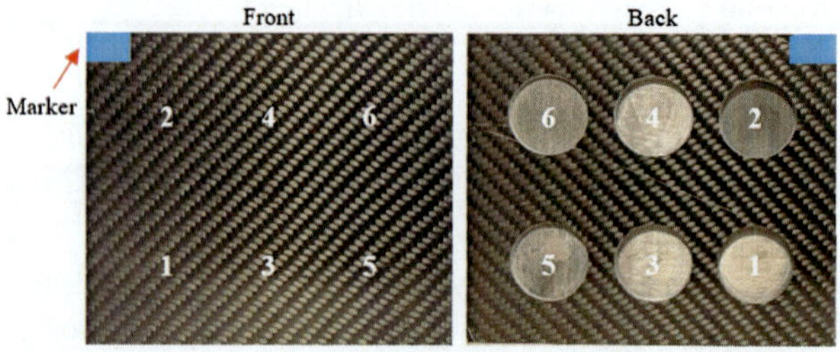

Defect No.	Perforation Depth*	Remaining Thickness*
1	4.00	1.50
2	3.50	2.00
3	3.00	2.50
4	2.50	3.00
5	2.00	3.50
6	1.50	4.00

****measurements in mm***

Fig. 1. CFRP sample details.

2.2 Experiment Setup

The experimental setup for the active IRT SC is shown in Fig. 2(a). The device used to record the IR radiation resulting from the thermal stimulation is a long-wave IR camera operating at a wavelength of 7.5–14 μm. The excitation source is a 650 W IR lamp. An excitation period of 10 s is used to heat up the CFRP sample.

3 Results and Discussion

3.1 Static Active IRT

The thermal image of the CFRP sample is seen in Fig. 2(b). The characteristic cooling curves representing the temporal temperature behavior of the respective defects and defect-free region derived from the static experiment are displayed in Fig. 2(c). The temperature difference (ΔT) between the temporal curves of the respective defects and the defect-free region are depicted in Fig. 2(d).

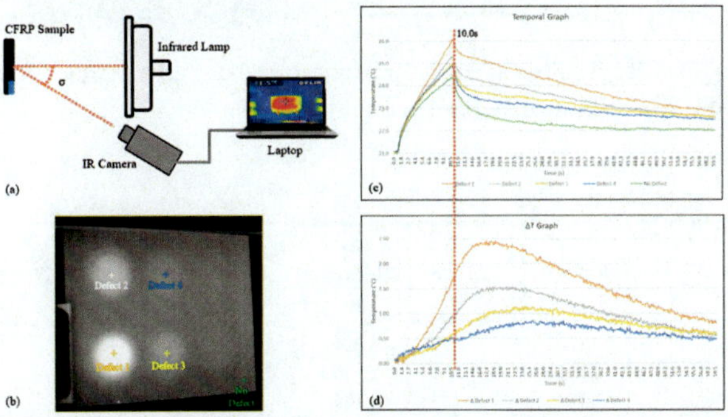

Fig. 2. Static active IRT results. (a) Static active IRT configuration. (b) Thermal image. (c) Temporal graph. (d) ΔT graph.

3.2 Dynamic Active IRT

As seen in Fig. 3(a), after the excitation duration of 10 s, the max ΔT for defects 1 and 4 occurs at 18 s and 26 s respectively. A gap of 8 s after excitation followed by an inspection window of 8 s (yellow region) is observed. The total inspection time is 26 s. After analyzing the SC data (*i.e.*, baseline data), the setup for the DC is derived as shown in Fig. 3(b). A conveyor belt with controllable speed is included in this setup. While the total inspection time is also 26 s, the DC requires the CFRP sample to move from the IR lamp to the IR camera within 8 s and start the IR camera recording within the inspection window of 18 s to 26 s. Altering the speed of the conveyor belt and/or the distance between the IR lamp and IR camera can yield similar results.

The SC requires the sample to be still for the entire duration of the inspection. However, in the DC, while one sample in being recorded; the IR lamp can simultaneously excite another sample. Thus, the DC provides the opportunity to inspect multiple identical samples at a faster pace compared to the SC. This is illustrated in Fig. 4 where a 20% timesaving is observed when inspecting three identical samples using the DC.

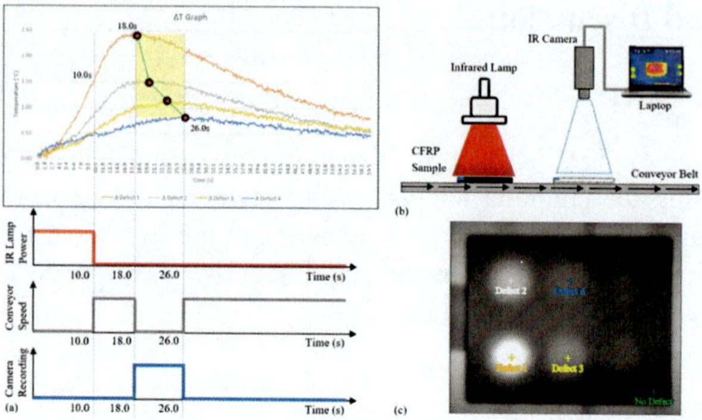

Fig. 3. Dynamic active IRT results. (a) Analysis of baseline data and dynamic active IRT sequence. (b) Dynamic active IRT configuration. (c) Thermal image.

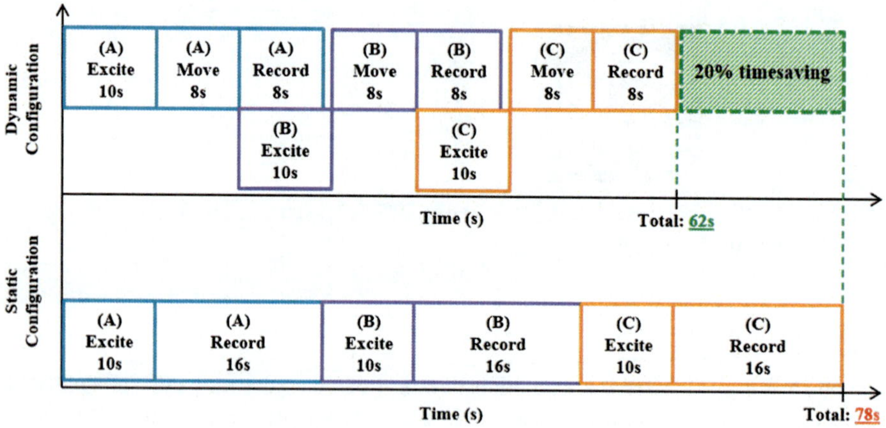

Fig. 4. Comparison of static and dynamic inspection times of three identical samples; A, B & C.

4 Conclusion

In this study, it is evident that both the static and dynamic active IRT inspection configurations provide similar temporal results. However, the DC has an advantage over the SC in terms of having a shorter overall inspection time when inspecting multiple identical components. With a suitable process workflow, the dynamic active IRT configuration allows for reduced inspection cycle times, cost-effective inspection process and has potential to be implemented in manufacturing/inspection lines.

Acknowledgements. This research work was supported by the Agency for Science, Technology and Research (A*STAR) through grants under its Singapore Aerospace Programme Cycle 13 (Grant No. A1915a0072) and Polymer Matrix Composites Programme (Grant No. A19C9a0044). Support of Singapore's National Research Foundation and the Building and Construction Authority is also acknowledged (Grant No. NRF2018NRF-CG002-005).

References

1. Maier, A., et al.: Non-destructive thermography analysis of impact damage on large-scale CFRP automotive parts. Materials (Basel, Switzerland) **7**(1), 413–429 (2014)
2. Peeters, J., et al.: Optimized dynamic line scanning thermography for aircraft structures. Quant. Infr. Therm. J. **16**(3–4), 260–275 (2019)
3. Busch, M., et al.: A robot inspection system allows the detection of defects in adhesive bonds between CFRP components by using active thermography, leading to reduces cycle times. In: 19th World Conference on Non-Destructive Testing (Munich, Germany) (2016)
4. Usamentiaga, R., et al.: Automated dynamic inspection using active infrared thermography. IEEE Trans. Industr. Inf. **14**, 5648–5657 (2018)

Determining Subsurface Defects' Depth in CFRP Composites Using Lock-In Thermography

Carlos Manzano(✉), Jonathan Zheng, Vinod Kumar, and Andrew Ngo

Institute of Materials Research and Engineering, 2 Fusionopolis Way, Innovis, #08-03, Singapore 138634, Singapore
{garciac,jonathan_zheng,vinod_kumar,ngocya}@imre.a-star.edu.sg

Abstract. The lock-in infrared thermography technique was used for inspecting a carbon fiber reinforced polymer (CFRP) sample to detect and to determine the depth of internal defects located underneath its top surface. The frequency dependency of incident modulated light on heat transport through the sample is explored to investigate its effect on the emitted thermal wave's phase and defects' visualization. An approach making use of localized phase shifts at defects visualized in phase images has been explored for determining the CFRP sample's defects depth.

Keywords: Infrared thermography · Lock-in · CFRP · NDT

1 Introduction

CRFP composites are being used to manufacture parts or components of airplanes [1], cars, motorbikes, high performance bicycles, etc. Composite components used in such critical applications, requiring to comply to stringent safety standards, undergo frequent inspections to ensure their structural integrity. There is a broad range of non-destructive techniques (NDT's) used to inspect composites e.g. X-Ray, ultrasonic, thermal, laser and terahertz sources, infrared sensors, etc. Among them, techniques based on active infrared thermography (IRT) are part of a group of NDT's that have being used and developed for detecting defects in a wide range of materials including composites [2]. IRT methods enable to detect and visualize sub-surface damages within composites, which cannot be detected by visual inspections, and to conduct large area inspections.

Lock-in thermography [3–5], an IRT technique, was used to inspect a CFRP sample with existing sub-surface defects with the aim of detecting the defects and to assess whether such technique can effectively be used for determining the defects' depth.

2 Methods

Lock-in thermal imaging is a technique implemented by heating a sample using infrared light sources whose intensity is modulated to conform with a sinusoidal waveform and

Y. Wei and S. Chng (Eds.): INCASE 2021, LNME, pp. 114–117, 2022.
https://doi.org/10.1007/978-981-16-5763-4_25

sensing the infrared thermal waves emitted from the sample using an infrared camera. Existing structural and compositional characteristics give rise to heat capacitance inhomogeneities within a sample. Such variations affect how thermal waves diffuse through and back propagate from a sample's structure resulting in changes in the amplitude and in phase shifts of thermal waves emitted by the sample relative to those of the incident modulated infrared light.

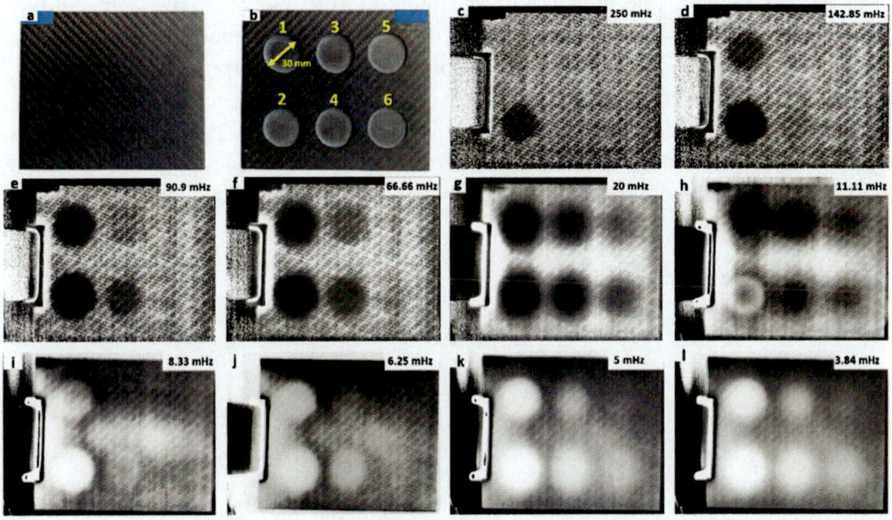

Fig. 1. a) CFRP sample's front, b) Sample's back showing circular defects of different depths. Depths with respect to the sample's top surface: D1: 3.93 mm, D2: 3.44 mm, D3: 2.95 mm, D4: 2.45 mm, D5: 1.94 mm, D6: 1.45 mm. c) to l) Selected characteristic phase images taken at frequencies ranging from 250 to 3.84 MHz. Deeper defects appear sequentially as dark circular areas while frequency is decreased. Images taken at lower frequencies e.g. below 20 MHz, show the defects appearing with reversed contrast, as bright circular areas.

One of the most important characteristics of lock-in thermography is the fact that there is an inversely proportional relationship between the frequency of the modulated incident light and the penetration depth of the heat waves diffusing through the sample. The experimental system used is made of a pair of infrared light sources (150W per lamp), one long wave infrared camera (FLIR A655SC), lock-in control unit and one laptop to control thermal image sequences' recording.

Experiments were done using a CFRP composite slab modified to create 6 circular cutouts of different depths on the sample's back. Visual inspection of the sample's front does not show any sign of the defects existing underneath, see Figs. 1a and b. Amplitude and phase images were taken at different frequencies ranging from to 250 MHz to 3.84 MHz. However, only phase images are relevant to this study. Phase images allow to visualize existing defects within a sample when they are acquired at frequencies low enough for thermal waves to diffuse through the sample to reach depths at which defects are located.

3 Results and Discussion

Phase images of the CFRP sample are shown in Fig. 1. They were taken at different frequencies ranging from 250 MHz to 3.84 MHz, only selected images are presented. Defects appear with different contrast with respect to defect-free areas i.e. sound areas, due to their different shifted phase values reflecting the structural or compositional variations across the imaged area.

In images taken at the high frequency range, 250 to 142.85 MHz, only the two shallowest defects are imaged, see Figs. 1c and d. As lower frequencies are used the other defects located deeper within the sample start to be visualized, see Figs. 1e to g. After all the defects were visualized the effect of even lower frequencies, below 20 MHz, was further investigated, see Figs. 1h to l. Those images show how as frequency is decreased the defects are visualized with inverted phase contrast, appearing as bright circular features, as compared with defects imaged at higher frequencies.

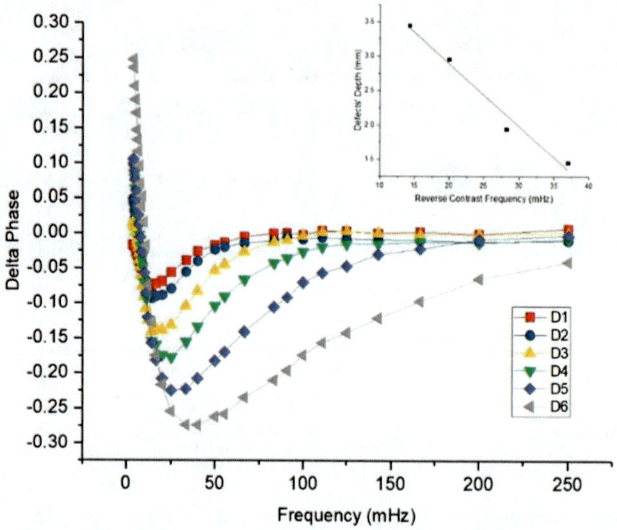

Fig. 2. Plot of defects' phase contrast changes, Δphase, with respect to the frequency at which images were taken. Inset: linear fitting to data of defects' depth with respect to reversal contrast frequency. Only measurements of four defects, D6, D5, D3 and D2, were used for the fitting.

The phase difference, Δphase, in between the phase at each defect and the sound area adjacent to it was computed to quantify how the defects' phase contrast changes with respect to the frequency at which the images were taken, see plots in Fig. 2. All plots present a similar shape with a characteristic skewered curve and a minimum at the low frequency range. Those curves track the phase contrast reversal from dark to bright observed in the phase images taken at low frequencies. The frequency minimum at each curve signals the threshold frequency at which the phase contrast reversal starts for the corresponding defect. Moreover, there is a clear inversely proportional relation in between the defects' depth and the contrast reversal onset, when the depth of the defects increases the threshold frequency shifts to lower frequency. As shown in the inset of Fig. 2, a linear expression can be approximated from the data points corresponding to threshold frequency and depth, and it could be used to estimate the depth of defects on CFRP composites. From that computed linear expression and corresponding phase contrast reversal threshold frequency the depth of the defect D4 is estimated to be 2.65 mm resulting in an overestimation of 8.28%. This estimation could be improved by increasing the experimental dataset in order to better determine the phase contrast reversal threshold for each defect.

4 Conclusion

Phase images taken using lock-in infrared thermography over a selected frequency range show how defects in a CFRP sample are imaged. Within a lower frequencies window defects' images show a phase contrast inversion that takes place at a frequency minimum for each defect and which depends on a defect's depth. An approach based on the relation in between defects' depth and the onset frequency of the contrast reversal was explored for estimating the depth of defects in CFRP samples. However, currently the method cannot determine defects' depth accurately enough and as such further research work needs to be done to improve it.

Acknowledgements. This research work was supported by the Agency for Science, Technology and Research (A*STAR) through grants under its Singapore Aerospace Programme Cycle 13 (Grant No. A1915a0072) and Polymer Matrix Composites Programme (Grant No. A19C9a0044). Support of the Singapore's National Research Foundation and Building and Construction Authority is also acknowledged (Grant No. NRF2018NRF-CG002-005).

References

1. Marsh, G.: Airbus A350 XWB update. Reinf. Plast. **54**, 20 (2010)
2. Ciampa, F., Mahmoodi, P., Pinto, F., Meo, M.: Recent advances in active infrared thermography for non-destructive testing of aerospace components. Sensors **18**, 609 (2018)
3. Busse, G., Wu, D., Karpen, W.: Thermal wave imaging with phase sensitive modulated thermography. J. Appl. Phys. **71**, 3962 (1992)
4. Bai, W., Wong, B.S.: Evaluation of defects in composite plates under convective environments using lock-in thermography. Meas. Sci. Technol. **12**, 142 (2001)
5. Liu, B., Zhang, H., Fernandes, H., Maldague, X.: Experimental evaluation of pulsed thermography, lock-in thermography and vibrothermography on foreign object defect (FOD) in CFRP. Sensors **16**, 743 (2016)

Surface Coatings

Electrochemistry and Engineering Study of Molecular Protective-Films to Steel-Alloys

Desmond T. C. Ang[1(✉)], Peter Junk[2], Glen Deacon[3], and Maria Forsyth[3]

[1] Singapore, Singapore
[2] College of Science and Engineering, James Cooks University, Townsville, QLD 4811, Australia
[3] School of Chemistry, Monash University, Clayton, VIC 3800, Australia
https://www.linkedin.com/in/t-c-desmond-ang-march672

Abstract. The syntheses and characterisation of Rare-Earth-Metals (REMs) complexes, via the use of green organic esters such as Pyridinecarboxylate (Pyc) and hydroxylnaphthoate (Hnp) to form stable compounds ie. Lanthanide-hydroxylnaphthanote (Ln-Hnp), has been demonstrated to be the most preferred compounds to form a good protective-film over steel-alloys. The electrochemical techniques of Linear Polarisation Resistance (LPR) and Cyclic-Potentiodynamic Polarisation (CPP) used in the studies demonstrated strong protective properties to the steel-alloys from the *E* corr-potential and *I* corr-current. This is because the *I* corr-current was reduced by many folds when coupons were immersed in NaCl (0.01M) as a control bulk-electrolyte. The data has also confirmed that the Ln-Hnp complexes clearly displayed a mixed-type of protective function. Thus, the overall metal dissolution rate of steel alloys is mitigated and the data also indicated that it has shown a healing mechanistic process building a dense molecular composite-films up to 5-micron thick when alloy-coupons were immersed into the solution containing REMs complexes. The characterisation of in-situ treated surfaces by the use of ATR-FTIR, FTIR-mapping-microscopy and Raman spectrum revealed its overall protective characteristic which were discussed in this paper.

Keywords: Corrosion inhibitors · Rare earth metals · Ligands · Coordination chemistry syntheses · Electrochemistry

1 Introduction

Redox reactions of metal-alloys with its environments is a natural dynamic-processes [1–9]. This works is working on Ln-complexes produced as green REMs-compounds, and to use for studying the ability on protection of steel-alloys. Characterisation of REMs-complexes and crystallography structures have been carried out. In summary, the formation of composite protective-film is expressed as:

D. T. C. Ang—On-Time Engineering Consultant.
P. Junk—Nevitt Prof. of Chemistry.

Y. Wei and S. Chng (Eds.): INCASE 2021, LNME, pp. 121–124, 2022.
https://doi.org/10.1007/978-981-16-5763-4_26

$$\left[Ln^{III}(L)_n\right]_{aq.} + y(OH^-) + x\left(Fe^{II}/Fe^{III}\right)_{aq} + x(L^-)_{aq.} + H_2O \rightarrow \left[\left(Fe^{II/III}\right)_x\left(Ln^{III}\right)_n\right](L)_{n+x-y}(OH)_y + Ln(OH)_{3(ppt.)} \quad (1)$$

2 Methods

The metathesis adopted to produce REMs complexes were strictly following the green principles. Reaction was to add REMs halides solution into ligand-salts with simple aqueous processing techniques in laboratory.

3 Results and Discussion of DOE Study on REMs-Ligands

The results of both the immersion-tests and *LPR* electrochemical data confirmed the lanthanide 3-hydroxy-2-naphthaote (3-Hnp) complex performs well in terms of the protective efficacy. This shows 3-Hnp-H is a relative stable poly-aromatic-hydrocarbon (PAH) compound, consisting of salicylato functional group i.e. carboxylate and ortho hydroxyl located at 2, 3 position of the core structure of naphthalene.

Table 1. Lab-Immersion test results *(ASTM G31-72)* [10] after seven days within the solutions of different chemical compositions.

($Ln^{(III)}$organic complexes) NaCl 0.10 M (100 mM) (zero additives) as control electrolyte	Dissolution rate (7d) (μg/m^2-s) Control 33.6 (±10%)	Factor of protective efficacy (%) ***
3-Hnp-Ln (where3-Hnp is 3-hydroxy-2-naphthoate Ln = **La; Ce; Pr; Mm** (Misch-metal ie. mixed REMs)	6.6; 4.5; 4.9; 4.3	81; 87; 85; 87
1-Hnp-Ln (where 1-Hnp is 1-hydroxy-2-naphthoate and Ln = **Ce**)	8.9	73
Isn-Ln (where Isn is isonicotinate and Ln = **La; Ce; Pr**)	15.5; 18.5; 18.5	46; 55; 56
Pic-Ln (where picolinate and Ln = **La; Ce; Pr**)	27.2; 15.1; 18.4	19; 46; 45

***Protective efficacy (%) defines as the reduction of alloy dissolution rate compared to control rate @ NaCl 0.10 M (100 mM) solution of 33.6 **μg/m^2−s.**

Table 2. ASTM G59-91 [11] Linear Polarisation Resistance (LPR) as R_p(ohm) measured and tabulated.

Inhibitors (Ln^{III} complexes)		*Rp* (K ohm) / E_{oc}(V)$_{(SCE)}$	Protective Efficacy % (***)
NaCl (10mM) **(zero additive)**		**1.9 / - 0.55 V**	**0**
3-Hnp-Ln	La (100 ppm/ 0.14 mM)	**24.8 / -0.45V**	**92.3**
	Ce (100 ppm/ 0.14 mM)	**25.3 / -0.38V**	**92.5**
	Pr (100 ppm/ 0.14 mM)	**47.3 / -0.39V**	**96.0**
	Mm (100 ppm/ 0.14 mM)	**28.5 / -0.38V**	**93.3**
Isn-Ln	La (200 ppm/ 0.41 mM)	**4.8 / -0.39V**	**60.4**
	Ce (200 ppm/ 0.40 mM)	**5.5 / -0.37V**	**56.4**
	Gd (200 ppm/ 0.39 mM)	**5.5 / -0.38V**	**56.4**
Pic-Ce	Ce (200 ppm/ 0.39 mM)	**3.2 / -0.31V**	**40.6**

(***) protective efficacy % is defined as increased of R_p *(ohm)* against the control R_p values.

The i_{corr} results of CPP plots (Fig. 1) for different Ln complexes, where Ln = La, Ce, Mm (mixed Ln) is in-line with the results of LPR data. It suggests that the combination of -OH and -COO groups at ortho proximity enables the formation of a strong binding coordination to the alloy surface of the laying metal substrate. There are research suggesting that a bimetallic compound forms with the Ce, Fe and the organic moiety.

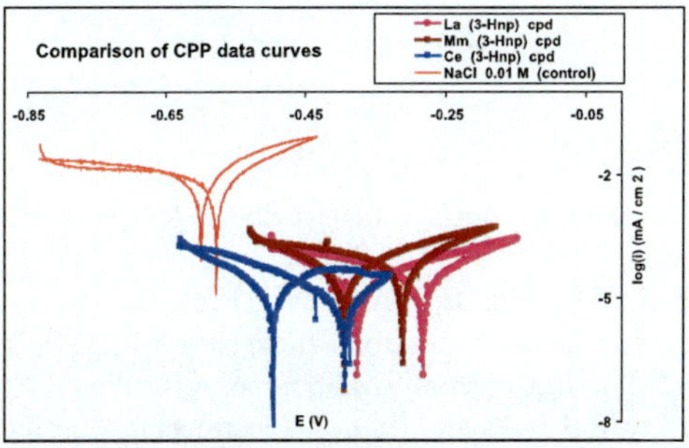

Fig. 1. CPP scan data of La, Ce and misch Metals (3-Hnp) complexes **La** (0.18 mM) **in pink, Ce** (0.14 mM) **in blue, Mm** (0.14 mM) **in brown** @ NaCl 0.01 M control solution (zero inhibitor) **in red**

3.1 Raman Spectrum of Ce 3-Hnp and AS1020 Steel-Alloy Coupons

The examination of in-situ surfaces by scanning with Raman 513 nm (green colour) on steel-alloy coupons, where coupons were treated within a bulk solution of NaCl 0.01 M containing Ce (3-Hnp) complex (0.14 mM). Raman-spectra indicated the formation of protective-film over steel-coupons, see Fig. 2. The result is in-line with the SEM images as seen from the EDAX-elemental analysis.

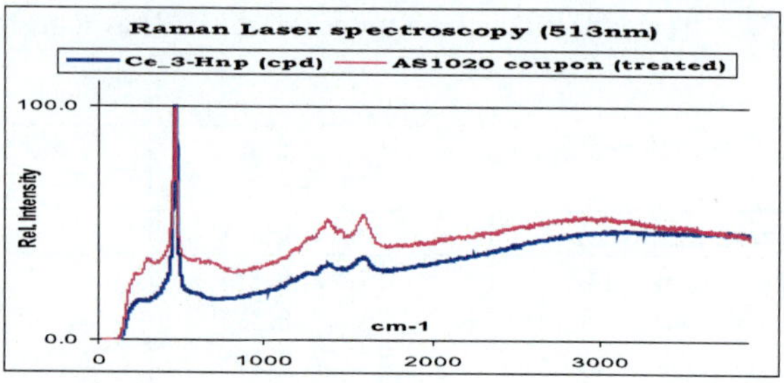

Fig. 2. In-situ surface study of steel-alloy with Raman spectroscopy (513 nm) @ control electrolytes containing Ce (3-Hnp) complex.

4 Conclusion

The results derived from the immersion-tests and *LPR/CPP* electrochemical data are presented in Table 1 and 2. We are confident that lanthanide 3-hydroxy-2-naphthaote (3-Hnp) performs well in terms of the protective efficacy for steel-alloy. The Ln (3-Hnp) complex has interestingly illustrated that it has a strong stability protective-film under the test with an aeration condition, and exhibiting an auto-healing effect which occurs in real-life application conditions.

Acknowledgement. The author greatly appreciates **Prof. Peter Junk, Prof. Glen Deacon and Prof. Maria Forsyth** for their guidance and assistance rendered during the entire work of my study.

References

1. Bockris, J.O.M., Khan, S.U.M.: Surface ElectroChemistry (A molecular Level approach). Plenum Press, New York (1993)
2. Tytgat, D., Solvay, B., et Cie.: (Ed.: E. P. Appl.) p. 21 French (1981)
3. Pludex, V.R.: Design and Corrosion Control. Wiley, New York (1997)
4. Blin, F.: Ph.D. thesis, Monash University, Melbourne, Australia (2005)
5. Hackerman, N., Snavely, E.S.J., Payne, J.S.J.: J. Electrochem. Soc. **113**(7), 677 (1966)
6. Leidheiser, H.J., Suzuki, I.: J. Electrochem. Soc. 128, 242 (1981)
7. Trabanelli, G., Carrassiti, V.: Advanced in Corrosion Science and Technology, vol. 1, p. 147. Plenum Press, New York (1970). (Eds.: M. G. Fontana, R. W. Staehle)
8. Evans, C.H.: Biochemistry of the Lanthanides. Plenum Press, New York (1990)
9. Kula, A.: J. Thermal Anal. Calorim. **68**, 957 (2002)
10. ASTM, in ASTM, G31-72 (Ed.: ASTM), p. 7 (1999)
11. ASTM, in ASTM, G59-91 (Ed.: ASTM), p. 4, Philadelphia, USA (1991)

Prediction of Microstructure Evolution of Cold Sprayed Coatings Using a Dislocation Density Based Constitutive Model

Sabeur Msolli[1](✉), Zhi-Qian Zhang[1], Debbie Hwee Leng Seng[2], Zheng Zhang[2], Junyan Guo[1], C. D. Reddy[1], N. Sridhar[1], Jisheng Pan[2], Boon Hee Tan[3], and Qizhong Loi[3]

[1] Institute of High Performance Computing, A*STAR Research Entities, #16-16, 1 Fusionopolis Way, Connexis, Singapore 138632, Singapore
{sabeur_msolli,zhangz}@ihpc.a-star.edu.sg
[2] Institute of Materials Research and Engineering, A*STAR Research Entities, #08-03, 2 Fusionopolis Way, Innovis, Singapore 138634, Singapore
[3] Advanced Remanufacturing and Technology Centre, A*STAR Research Entities, #01-01, 3 Cleantech Loop, CleanTech Two, Singapore 637143, Singapore

Abstract. A dislocation density based model is implemented as a user hardening subroutine in Abaqus within the Eulerian finite element framework to predict the Geometrically Necessary Dislocations (GND) density and grain refinement during the cold spray deposition process of Al6061-T6. The model is capable of exploring the dependency of the microstructure on the cold spray process parameters such as particle velocity, deposition angle, initial temperature and particle size. Furthermore, the model depicts microstructural evolution in critical regions such as the coating-substrate interface and particle boundary and provides insights into the microstructural distribution observed at different locations in a cold sprayed part.

Keywords: Cold spray · Dislocation density · Cell size · Mis-orientation angle · Grain refinement

1 Introduction

Metal cold spray consists of propelling metal powder onto a substrate for realizing a protective coating layer. Upon hitting and depositing onto the substrate, the powder particles have their microstructure change drastically due to the extreme temperature increase originating from severe plastic deformation. In the microscale, these changes consist of variations in dislocation movement related to the metamorphosis of the grain structure transforming into smaller subgrains. At the end, the mechanical properties of the coating are strongly dependent on many factors including the particle velocity and deposition angle, the initial microstructure of the powder particle and the substrate, the preheating temperature among others [1]. The aim of the present paper is to predict the the evolution of the geometrically necessary density in the as-sprayed cold spray coating.

Y. Wei and S. Chng (Eds.): INCASE 2021, LNME, pp. 125–128, 2022.
https://doi.org/10.1007/978-981-16-5763-4_27

We use a dislocation density based constitutive model proposed by Estrin et al. [2], that captures the hardening stages as well as the changes in grain size, misorientation angle and dislocation density during and after the deposition. The model is able to describe the strain rate and temperature dependency of the material making it very suitable for modelling such extreme manufacturing processes. This dislocation density-based model is implemented as a user subroutine in the commercial FEM software package Abaqus to handle multiple particle impact as in the cold spray process. Validation of the results are carried out with experiments measuring the GND density.

2 Methods

Al6061 powders were purchased from Plasma Giken Co Ltd (part number PG-AMP-1170) and pre-heated at 60 °C for 1 h prior to spray. The mean particle size is 60 μm. There was no sand blasting on the Al6061-T6 coupons, but they were exposed to the coating powder in their as-received state. The cold spray process was performed using Plasma Giken's PCS-1000 system with pre-heated nitrogen gas. The whole procedure consisted of 5 deposition passes to obtain a coating of 146 μm in thickness. The coating was obtained using a nominal particle velocity of 625 m/s and a nozzle impingement angle of 90°. Experimental measurements were carried out to characterize some of the Al6061-T6 coated assemblies. In these experiments, SEM and EBSD were used to observe the microstructure generated in the coating and the coating-substrate interface.

Support of the experiments was ensured with finite element simulations of the cold spray process using the dislocation density material model that describes the deformation and microstructure changes including dislocation density, misorientation and grain refinement [2, 3]. The finite element model of the cold spray process consists of over 2000 particles impacting a substrate with a particle velocity of 625 m/s and impact angle of 90°, consistent with experiments. All substrate faces are assumed to be non-reflecting boundaries except the one subjected to the coating process which is assumed to be a free surface. The simulations are thermos-mechanically coupled considering the heat produced by severe plastic deformation and the heat conduction that dissipates the generated heat in the whole assembly. The substrate considered was large enough to inhibit the effect of the boundaries on the simulation results. The results of interest are the GND density and the grain size evolution which are volume averaged over a suitable domain and analyse at intervals of 5 μm to capture the variation of these variables in space.

3 Results and Discussion

The experimental observations using SEM show a high density coating with a crack at the coating substrate interface and a few distributed voids as shown in Fig. 1a. The band contrast using EBSD showed particles severely flattened. Nano-crystalline layers are also displayed as non-indexed black areas which are aggregated near the particle boundaries (Fig. 1b). Particles' boundaries are locations where high plastic shear deformation and extreme heating likely occur [4], leading to recrystallization and grain size modification. KAM images in Fig. 1c show higher GND density at the coating and the substrate top surface owing to the severe plastic deformation at these regions.

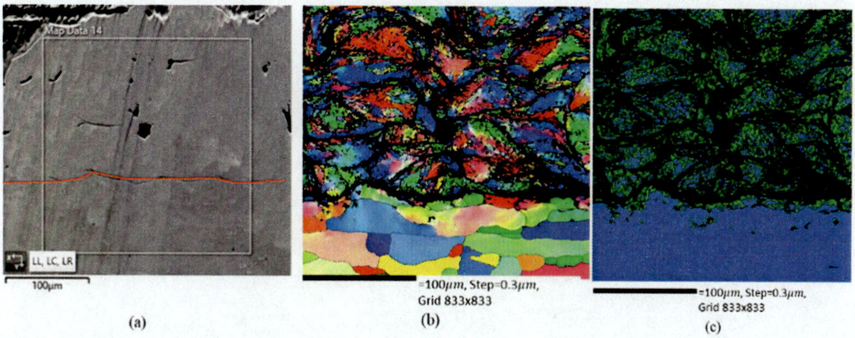

Fig. 1. (a) Forward scattered electron image of coating-substrate interface region; (b) EBSD band contrast and (c) orientation map of the area highlighted in white box in (a).

For comparison, we show the coating microstructure predicted by the finite element simulations in Fig. 2. GND density and misorientation are significant at the particle boundaries as shown in Fig. 2(a, b) and at the substrate top layer. As the cell size is directly related to the GND density, intense grain refinement occurs at these specific locations (Fig. 2c) corresponding to the dark domains observed using EBSD (see Fig. 1b, for instance). KAM images of the coating have been used to compare the measured and calculated GND density along the coating.

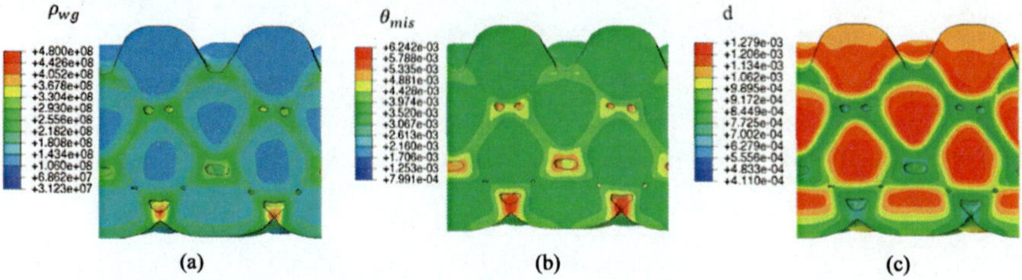

Fig. 2. Microstructure maps in coating. (a) GND dislocation density (unit: mm-2), (b) misorientation angle (unit: rad), (c) cell size (unit: mm).

Figure 3a illustrates the zone of interest where the GND density is extracted. As seen in Fig. 3b, a good prediction of the GND density across the coating is achieved using the dislocation density model. A peak in GND density is observed at the interface of the coating and substrate where the particles are the mostly deformed and flattened duc to high velocity impact. It stabilizes immediately to its final value of 7×10^8 mm^{-2}.

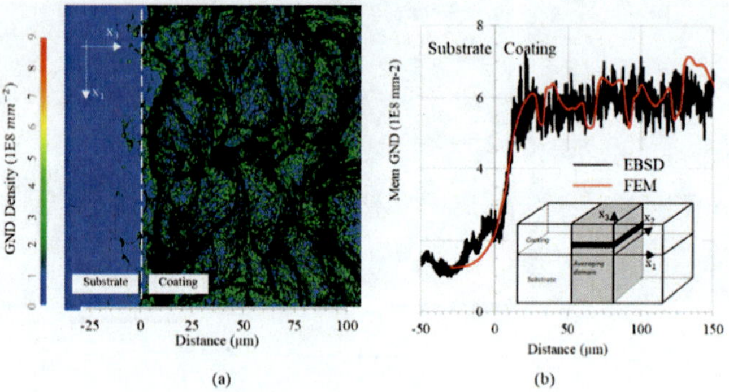

Fig. 3. (a) Variation of the GND density in the coating and the coupon. (b) FEM and experimental Mean GND distribution in the coupon/coating interface.

4 Conclusion

A dislocation density based model is implemented to evaluate the microstructural changes for cold sprayed Al6061-T6 coatings. The experimental data for the microstructure of the coating and finite element simulations of the cold spray process were compared for the GND density. We have good agreement between the computed and measured GND density. The dislocation density based model seems a good framework in predicting microstructural evolution under extreme conditions, especially as in the cold spray process.

Acknowledgment. This research is supported by the Agency for Science, Technology and Research (A*STAR) under the program "Machining Learning Assisted Control of Metal Cold Spray and Shot Peening Processes (A1894a0032)" and Singapore Aerospace Program (A1715a0073).

References

1. Chen, C., et al.: Cold spraying of thermally softened Ni-coated FeSiAl composite powder: Microstructure characterization, tribological performance and magnetic property. Mater. Des. **160**, 270–283 (2018). https://doi.org/10.1016/j.matdes.2018.09.025
2. Lemiale, V., Estrin, Y., Kim, H.S., O'Donnell, R.: Forming nanocrystalline structures in metal particle impact. Metall. Mater. Trans. A **42**(10), 3006–3012 (2011). https://doi.org/10.1007/s11661-010-0588-5
3. Tóth, L.S., Estrin, Y., Lapovok, R., Gu, C.: A model of grain fragmentation based on lattice curvature. Acta Mater. **58**(5), 1782–1794 (2010). https://doi.org/10.1016/j.actamat.2009.11.020
4. Reddy, C.D., Zhang, Z.-Q., Msolli, S., Guo, J., Sridhar, N.: Impact induced metallurgical and mechanical interlocking in metals. Comput. Mater. Sci. **192**, 110363 (2021). https://doi.org/10.1016/j.commatsci.2021.110363

Parameters Optimization for Laser Cladding Fe55 Alloy on 40Cr Steel

Cheng Li, Yanle Li(✉), Xiaoxia Qi, Jiyu Du, and Fangyi Li

School of Mechanical Engineering, Shandong University, Jinan 250061, China
yanle.li@sdu.edu.cn

Abstract. In order to explore the process parameters of laser cladding Fe55 alloy on 40Cr steel, a three-factor, three-level orthogonal experiments were conducted. The influence law of laser power, scanning speed and powder feed rate on the dilution rate and microhardness of the cladding layer are explored. The order of the influence of factors on the dilution rate from greatest to least was determined to be as follows: powder feed rate > scanning speed > laser power. The order of the influence of factors on hardness is scanning speed > powder feed rate > laser power. The optimal parameters to obtain the largest hardness were with a powder feed rate of 0.8 rpm, a laser power of 1300 W, and a scanning speed of 900 mm/min.

Keywords: Laser cladding · Process parameters · Orthogonal experiment

1 Introduction

Laser cladding [1] is a surface modification technology in which powder alloy is rapidly heated and melted on the surface of the substrate under the action of a high power laser beam to produce a cladding layer with fine structure, low porosity, and metallurgical bonding with the substrate material. It can be used to prepare surface strengthening layers with various characteristics such as wear resistance, corrosion resistance, high temperature resistance, and oxidation resistance, etc. [2].

In recent years, there have been many studies on single-channel and single-layer laser cladding on 40Cr surface. Wu et al. [3] used laser cladding single-pass Fe-based coatings on 40Cr substrates,the hardness of the cladding layer obtained was 2.9 times that of the substrate, and the wear amount is only 54%. Ren et al. [4] cladded iron-based alloys on the surface of 40Cr steel, which shows a good surface morphology and excellent friction properties. Mao et al. [5] cladded iron-based materials on 20CrNiMo steel to improve its wear resistance. For parts with severe wear, a single-layer cladding coating cannot meet the actual needs. Based on orthogonal experiments, the present

Y. Wei and S. Chng (Eds.): INCASE 2021, LNME, pp. 129–133, 2022.
https://doi.org/10.1007/978-981-16-5763-4_28

work explores the influence of laser power, scanning speed and powder feed rate on the dilution rate and hardness during the double-layer multi-channel cladding. This work lay a good foundation for Fe55 alloy repairing 40Cr matrix in the future.

2 Methods

2.1 Materials

The cladding substrate is a 40Cr steel plate, with a size of 50 mm × 50 mm × 10 mm, and all six faces are milled into a smooth surface, so that the flatness is guaranteed to be within 0.04 mm. The powder used is an iron-based Fe55 self-fluxing alloy, with the size of 50–100 μm, and the specific chemical composition of the matrix and powder materials is shown in Table 1.

Table 1. Chemical composition of experimental materials (wt.%)

Material	C	Si	Mn	Cr	Ni	Mo	P	S	B	Fe
40Cr	0.4	0.27	0.6	1	≤0.25	–	≤0.03	–	–	Bal.
Fe55	1.6	2.4	–	26.4	11.2	–	–	–	1.7	Bal.

2.2 Equipment and Methods

A 2 kW YCWL-HP-P1500 fiber laser was used for laser cladding, which is shown in Fig. 1. The beam spot is round, and the energy presents a nearly Gaussian distribution. Besides, Ar were chosen as the carrier and protective gas. The sprayed plate was cut along the direction perpendicular to the laser scanning. After mounting and polishing the sample, a 4% nitric acid-alcohol etching solution was used for metallographic etching. The cross-sectional photos were characterized using the JSM-7610F Schottky field emission scanning electron microscope. The microhardness was measured using a Vickers hardness tester at a load of 200 g and pressure holding time of 15 s. The cladding height *H*, cladding depth *h* were measured by Image J software. A series of tests is designed with the orthogonal experiment method, as shown in Table 2. Three factors including laser power P, scanning speed V_s and powder feed rate V_f are selected and each factor is defined with three levels.

Fig. 1. Laser cladding equipment

Table 2. Orthogonal experimental parameters

Sample	Power (w)	Scanning Speed (mm/min)	Powder feed rate (rpm)
1#	1100	500	0.8
2#	1100	700	1.2
3#	1100	900	1
4#	1300	700	1
5#	1300	900	0.8
6#	1300	500	1.2
7#	1500	900	1.2
8#	1500	500	1
9#	1500	700	0.8

3 Results and Discussion

Dilution rate refers to the degree to which the substrate material melts into the cladding layer during the cladding process, which influences the composition directly. If the dilution rate is too large, more impurities enter the cladding layer, which reduces the performance of the cladding layer. In the case of a small dilution rate, it is difficult to form a strong metallurgical bond between the cladding layer and the substrate [6]. The dilution rate can be calculated by formula (1), and the parameters of the formula are shown in Fig. 2.

$$\eta = \frac{h}{h + H} \times 100\% \tag{1}$$

where η is the dilution rate, h is the cladding depth (mm), and H is the cladding height (mm).

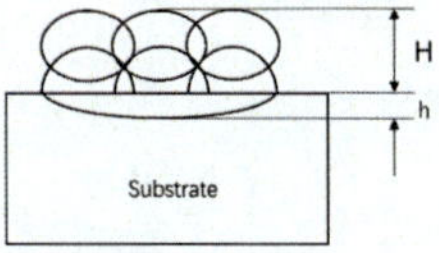

Fig. 2. Section size of cladding layer

The calculated dilution rates η are provided in Table 3. The hardness value curve of each point measured is shown in Fig. 3. The range analysis results of dilution rate and hardness are recorded in Table 4.

Table 3. Values of cladding height, cladding depth and dilution rate

Sample	*H* (mm)	*h* (mm)	η(%)
1#	2.27	0.68	0.23
2#	1.96	0.45	0.19
3#	1.33	0.53	0.29
4#	1.87	0.55	0.23
5#	1.30	1.65	0.33
6#	3.33	0.63	0.16
7#	1.84	0.69	0.27
8#	3.11	0.83	0.21
9#	1.55	0.84	0.35

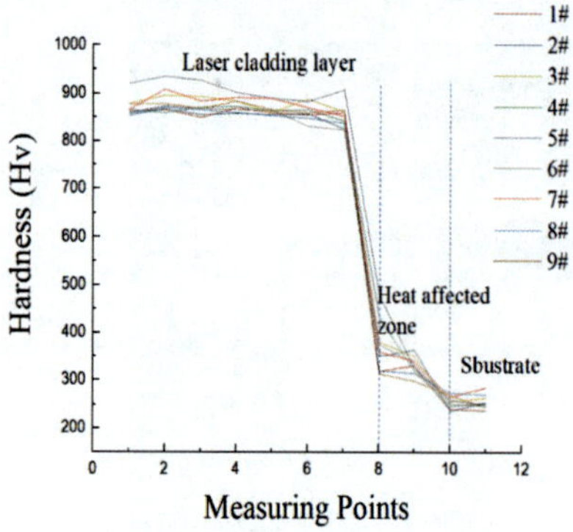

Fig. 3. Microhardness distribution

According to data of the dilution rate η in Table 4, it can be seen that R (V_f) > R (V_s) > R (P). It is suggested that the most important factor affecting the dilution rate is the powder feed rate, followed by the scanning speed, and finally the laser power. This is because that with the speed of the powder feed rate increases, more laser energy is needed to melt the powder, so the energy to melt the substrate is correspondingly reduced. In addition, the increase of powder causes the diffuse scattering of laser radiation [7], resulting in the deficiency in energy reaching the matrix. Thereby reducing the dilution rate.

According to the hardness data in Table 4, the biggest influence on the hardness is the scanning speed, then the powder feed rate and finally the laser power. The high scanning speed increases the cooling rate and shorten the growth time of the structure, thereby inhibiting the growth of dendrites, leading to grain refinement. Grain refinement increases the number and area of grain boundaries, thereby improving the performance

Table 4. Results of dilution rate and hardness range analysis

Index		P (w)	V_s (mm/min)	V_f (rpm)
Dilution rate (%)	K1	0.23	0.20	0.30
	K2	0.24	0.25	0.24
	K3	0.27	0.29	0.20
	R	0.04	0.09	0.10
Hardness (Hv)	K1	867.00	856.35	881.06
	K2	876.04	862.45	869.18
	K3	869.16	896.40	861.96
	R	9.04	37.05	19.1

of the cladding layer. The best process parameters for maximum hardness obtained as: laser power 1300 W, scanning speed 900 mm/min, and powder feed rate 0.8 rpm.

4 Conclusion

Through orthogonal experiments, the influence of laser power, scanning speed and powder feed rate on the dilution rate and hardness of the cladding of Fe55 alloy on the surface of 40Cr steel are explored. The following conclusions can be drawn:

(1) As for the dilution rate, the most important factor is the speed of the powder disc, followed by the scanning speed, and finally the laser power. (2) The order of influence on hardness is the scanning speed, then the rotating speed of the powder disc, and finally the laser power. Within the tested range of experiments, the best process parameters can be obtained: laser power 1300 W, scanning speed 900 mm/min, powder feed rate is 0.8 rpm.

References

1. Pinkerton, A.J.: Lasers in additive manufacturing. Opt. Laser Technol. **78**, 25–32 (2016)
2. Singh, S., Goyal, D.K., Kumar, P., Bansal, A.: Laser cladding technique for erosive wear applications: a review. Mater. Res. Express **7**(1), 020 (2007)
3. Wu, P., Lei, S., Li, S., Zhu, J., Liang, D., Gu, C.: Study on the structure and properties of laser cladding coating on the surface of 40Cr roll. J. Heilongjiang Inst. Technol. (Comprehensive Edition) **20**(08), 47–52 (2020)
4. Ren, J., Yang, X., Huang, H., Zheng, L.: Microstructure and performance analysis of laser cladding layer of 40Cr die steel. Die Mould Ind. **43**(07), 59–61+64 (2017)
5. Mao, J., Feng, A., Cheng, B., Li, Z.: Microstructure and friction and wear properties of fe-based alloy coatings by laser cladding. Hot Work. Technol. **46**(02), 139–142 (2017)
6. Li, J.N.: Laser Cladding Technology and Application. Chemical Industry Press, Beijing (2015)
7. Alvarez, P., Montealegre, M., Pulido-Jiménez, J.: Analysis of the process parameter influencein laser cladding of 316L stainless steel. J. Manuf. Mater. Process. **2**, 55 (2018)

Feasibility Tests on Directed Laser Deposition and Cold Spray Coatings Applied in Series for Different Industrial Applications

Davide Verdi(✉), Qizhong Loi, Boon Hee Tan, and Alin Patran

Agency for Science, Technology and Research (A*STAR),
Advanced Remanufacturing and Technology Centre (ARTC), 3 CleanTech Loop,
CleanTech Two, #01/01, Singapore 637143, Singapore
{davide_verdi,loiq,tanbh,alin_patran}@artc.a-star.edu.sg

Abstract. Directed Laser Deposition and Cold Spray are commercially available techniques used to apply metallic and metal matrix composites coatings for different functions, such as wear and corrosion resistance. Even if these methods are already certified in different industrial sectors, their more extensive use is sometimes restricted by technical limitations. For example, the Direct Laser Deposition on Aluminum is challenging due to the high reflectivity and low absorption of near infrared light of this material. On the other hand, the corrosion resistance achievable through Cold Spray is sometimes limited due to the intrinsic porosity of the coatings applied with this technique. In this paper, the results obtained from feasibility tests on the synergy of the two methods is reported. In particular, Inconel 718 Cold Spray coatings were applied on Aluminum coupons as buffer layer for different Directed Laser Deposition coatings. On the other hand, fully dense Inconel 625 laser deposited coatings were applied as barrier against corrosion on steel substrates and then coated via Cold Spray. A preliminary analysis of the microstructure of the obtained coatings was studied. The benefits achievable with this hybrid approach were also discussed.

Keywords: Directed Laser Deposition · Cold Spray · Aluminum · Wear resistance · Corrosion resistance

1 Introduction

Directed Laser Deposition (DLD) is an Additive Manufacturing (AM) process in which a focused laser beam is used to create a meltpool on the workpiece surface where a feedstock material, in form of powder or wire, is injected. As the laser is moving following a determined toolpath, a solid weld bead is formed characterized by a strong metallurgical bond with the underlaying material. Through DLD it is possible to manufacture near net shape 3D parts, repair damaged components, or apply coatings for different functional applications [1]. Cold Spray (CS) is a solid-state coating process in which a feedstock powder is accelerated by a preheated gas through a de Laval converging-diverging nozzle. The powder exiting the nozzle remain in a solid state and reach velocities of the

Y. Wei and S. Chng (Eds.): INCASE 2021, LNME, pp. 134–137, 2022.
https://doi.org/10.1007/978-981-16-5763-4_29

order of 600–1000 m/s. The powder particles experiment a plastic deformation when they hit the workpiece surface and a mechanical bond is generated. In this way, metallic coatings made of layers of hundreds of microns in thickness can be manufactured [2].

Some technical limitations may however affect the deployment of DLD and CS in different industrial fields. For example, in DLD mainly lasers with near infrared wavelengths ($\lambda \sim 1\ \mu m$) are employed, but they are difficult to use on highly reflective materials such as Aluminum, Copper, or Silver [3]. On the other hand, even if CS represents an ideal solution to apply hard and wear resistant coatings, its intrinsic level of porosity limits its application, for example, in corrosive environments.

By coupling DLD and CS, it will be possible to overcome their limitations and extend their applications in more industrial fields. For example, a CS coating can be used as buffer layer to allow the deposition of coatings of different specific properties (e.g. mechanical, functional) on Aluminum parts. Then again, both corrosion and wear resistance of a specific component can be enhanced by the application via DLD of a fully dense layer between substrate and CS coating. In this paper, the results of some preliminary tests on the usage in series of the two technologies are reported. While the use in tandem of lasers and CS was reported in literature, for example, to improve the adhesion of the CS coatings [4], to the authors' knowledge, the use of DLD and CS in series was not reported or described before.

2 Methods

To deposit the DLD coatings a commercial CNC machine equipped with a fiber laser ($\lambda = 1{,}070\ \mu m$) with a maximum output power of 2 kW and a focused spot size of 2.2 mm was used. The feedstock material was delivered to the meltpool generated by the laser beam through a coaxial powder feed nozzle. Argon was used as powder carrier and process shielding gas. The CS coatings were applied through a commercial system with maximum chamber gas temperature of 1000 °C and maximum gas pressure of 5 MPa. Nitrogen was used as process gas in this case.

Ten millimeters thick AA6061 Al-alloy coupons were used as substrates and a 0.5 mm thick Inconel 718 (In718) coating was applied via CS to work as buffer layer for DLD. Three materials were selected to be tested via DLD: Inconel 625 (In625), Copper 99.3+% (Cu), and Titanium Carbide (TiC). One-layer patches of $20 \times 20\ mm^2$ were deposited following a zig-zag toolpath. On the other hand, six millimeters thick stainless steel 304 (SS304) coupons were used as substrates and In625 coatings of approximatively 1 mm thickness were applied on them via DLD. An In718 coating was then applied on top through CS. In both cases, the parameters selected for the deposition of these materials were selected based on previous experience.

The coatings were then sectioned, mounted in resin, and polished. Their cross sections were finally analyzed by means of an optical microscope.

3 Results and Discussion

Optical images of the In625, Cu, and TiC DLD coatings deposited onto In718 CS layers are reported in Fig. 1, Fig. 2, and Fig. 3 respectively. As can be seen, the laser did not penetrate completely the CS layer and therefore did not reach the AA6061 substrate. While the adhesions between DLD and CS layers, as well as between the CS layers and substrate look proper without defects, a crack can be observed within the CS layers in all cases, more evident where the first DLD track was deposited. Despite this, no delamination of the coatings was experienced. The cracks seem to be formed at the borders of the heat affected zones (HAZ) generated by each DLD track, most probably due to the microstructural changes and stress release imposed by this process.

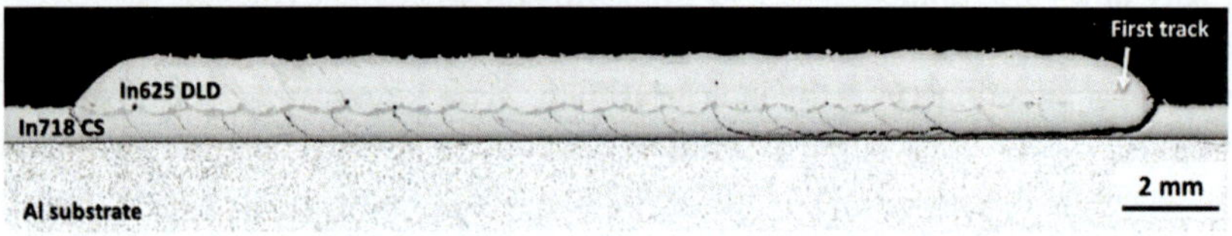

Fig. 1. In625 DLD coating on AA6061 substrate with In718 CS buffer layer

Fig. 2. Cu DLD coating on AA6061 substrate with In718 CS buffer layer

Fig. 3. TiC DLD coating on AA6061 substrate with In718 CS buffer layer

One possible way to eliminate these undesired defects can be to reduce the thickness of the CS layers to allow their full remelting and some degree of dilution with the aluminum substrate. However, this can lead to the formation of undesired phases that can alter the properties of the coating. Alternatively, a pre-DLD process heat treatment can be carried out to reduce the intrinsic residual stresses in the CS layer. These and other solutions will be further investigated to address properly this issue and achieve crack-free coatings.

In Fig. 4 it is possible to appreciate the In718 CS coating applied onto the In625 DLD layer. A dashed line was added to highlight the interface of the DLD coating with the SS304 substrate, free of defects like cracks, pores, or inclusions. On the other hand, at the CS/DLD interface it is possible to appreciate the presence of darker areas corresponding to cracks and lack in adhesion. This issue can be solved by a proper pre-CS deposition preparation of the DLD layer surface, such as removal of the surface oxides and waviness (peaks and valleys) of the DLD build-up. This, as well as the use of different materials combinations, will be investigated further to identify possible limitations and propose solutions for wear resistant coatings in corrosive environments.

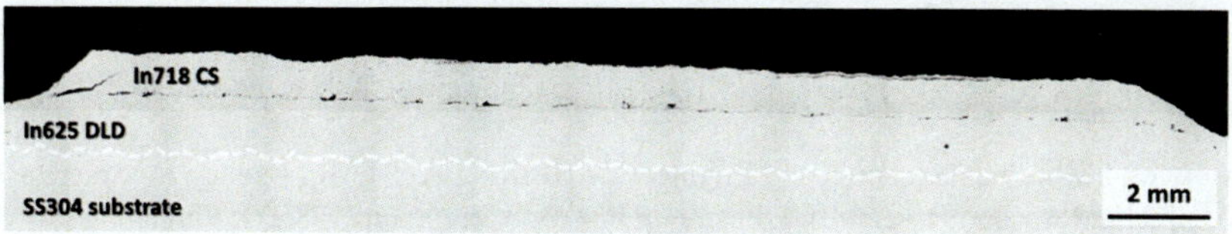

Fig. 4. In718 CS coating applied on In625 DLD layer deposited on SS304 substrate

4 Conclusions

Preliminary tests on the use of DLD and CS in series to apply coatings able to overcome some of the limitations of each of these two processes were presented. The CS buffer layer allowed the use of a near infrared laser to apply DLD coatings onto an aluminum substrate. However, cracks were formed within the CS coating, possibly due to the heat originating from the DLD process. On the other hand, when an In718 CS coating was applied on an In625 DLD layer, defects were observed at their interface. A proper surface preparation of the DLD layer before CS could solve this issue. Despite the obtained coatings always presented defects such as cracks or lack in adhesion, these first tests demonstrated the feasibility of using these two technologies in series to extend their industrial applications. Further tests and analysis will be carried out to completely understand the reported issues and achieve defects-free coatings.

Acknowledgements. The authors would like to acknowledge A*STAR and ARTC for the financial support.

References

1. Shamsaei, N., Yadollahi, A., Bian, L., Thompson, S.M.: An overview of Direct Laser Deposition for additive manufacturing; Part II: mechanical behavior, process parameter optimization and control. Addit. Manuf. **8**, 12–35 (2015). https://doi.org/10.1016/j.addma.2015.07.002
2. Dykhuizen, R.C., Smith, M.F.: Gas dynamic principles of cold spray. J. Therm. Spray Technol. **7**, 205–212 (1998). https://doi.org/10.1361/105996398770350945
3. Indhu, R., Vivek, V., Sarathkumar, L., Bharatish, A., Soundarapandian, S.: Overview of laser absorptivity measurement techniques for material processing. Lasers Manuf. Mater. Process. **5**(4), 458–481 (2018). https://doi.org/10.1007/s40516-018-0075-1
4. Barton, D.J., Hornbuckle, B.C., Darling, K.A., Brewer, L.N., Thompson, G.B.: Influence of surface temperature in the laser assisted cold spray deposition of sequential oxide dispersion strengthened layers: microstructure and hardness. Mater. Sci. Eng. A. **811**, 141027 (2021). https://doi.org/10.1016/j.msea.2021.141027

Advanced Surface Engineering and Protective Coating

Hongfei Liu(✉), Tzee Luai Meng, Jing Cao, and Chee Kiang Ivan Tan

Institute of Materials Research and Engineering (IMRE), A*STAR (Agency for Science, Technology and Research), 2 Fusionopolis Way, Singapore 138634, Singapore
liuhf@imre.a-star.edu.sg

Abstract. The surface and near-surface layers of structural metal alloys are the front line against corrosion, erosion, and wear attacks from their working environments. To enhance the surface layers without degrading the mechanical strength of the base material, various surface engineering technologies have been redeveloped in the past decade driven by ever-increasing requirements for additive manufacturing of metallic components. In general, the surface engineering techniques can be catalogued into two groups. One is surface treatment that either enhances the intrinsic surface materials via introducing plastic deformations or converts them into more thermodynamically stable materials upon direct energy flux. The other is surface coating by single or multiple layered extrinsic materials that have higher corrosion and wear resistances than those of the base material. Hybrid methods between the two groups have also been proposed in recent years for advanced surface engineering. Here, we briefly introduce a few techniques for advanced surface engineering and protective coating, highlighting their remaining challenges and promising potentials.

Keywords: Surface engineering · Cold-working · Protective coating

1 Introduction

Research and development in additive manufacturing (AM) and remanufacturing of metallic components have been rapidly growing in the past decades along with the maturing of industrial digitalization and automation technologies [1]. Although some AM techniques have been employed in practical repair and maintenance operations, post-AM processing are often necessary due to the metastable solidification and the rough surface of the AM materials [2]. As a consequence, advanced surface enhancements, e.g., cold-working, laser treatment, etc., have been redeveloping, either integrated in AM or separately processed after AM, to improve the integrity of the AM components are discussed.

Coating technologies, such as thermal spray, cold spray, arc spray, laser-cladding, etc., are typical AM processes from the point of view of materials deposition and building-up. They provide extrinsic overlay in single or multiple layers with desired surface performances prior to those of the substrate, e.g., against corrosion, erosion, and wear in aggressive corrosion environment. Both the coating material and the coating process play an important role in developing advanced coatings towards desired purpose.

Y. Wei and S. Chng (Eds.): INCASE 2021, LNME, pp. 138–141, 2022.
https://doi.org/10.1007/978-981-16-5763-4_30

Precisely identifying the relationships amongst the source material, the coating process, and the resultant metallurgical structures is helpful in engineering the performance of the coating layer with improved cost-efficiency.

In this article, we briefly introduce advanced surface engineering and protective coating, taking robotic hammer peening (RHP), severe shot peening (SSP), laser-treatment, laser-cladding, and twin-wire arc spray (TWAS), as typical examples. Their remaining challenges and promising potentials for industrial fabrications as well as coating protections of structural alloys towards sustainable infrastructures.

2 Surface Engineering

Surface engineering is a branch of materials science and technology, its coverage varies for different areas from chemistry, metals, ceramics, to semiconductors. Cold-working and laser-treatment have been extensively developed in metal-related fabrications to modify the surface towards enhanced integrity and hence the fatigue life of the components [3, 4].

2.1 Cold-Working

Physically, cold-working introduces plastic deformation, resulting in surface hardening and compressive residual stress in the surface layer of the workpiece. There are many cold-working processes such as peening, rolling, burnishing, etc. We have recently studied RHP and SSP [5–7]. In general, severe cold-working may induce grain refinements due to the accumulation and interaction of densified dislocation networks [7]. Grain refinements tend to improve the surface hardness; they may also improve the corrosion resistance of the workpiece, e.g., via mitigating the pitting initiations as well as the cracking propagations. Nevertheless, the effect of cold-working on corrosion resistance of the workpiece is greatly dependent on the process as well as on the material of the workpiece. Plastic deformation induced martensite transformations could be caused by cold-working in stainless steels with relatively low stacking fault energies (e.g., SS304). The effect of the coexisted austenite and martensite grains on the corrosion resistance of the workpiece is still under debate due to the complexed structural modifications in tandem with the phase transformations [4, 6]. Electrochemical methods have been widely employed to evaluate the corrosion resistance influenced by cold-working; however, the surface roughness has to be carefully taken into considerations when comparing the corrosion resistance between the workpiece experienced different cold-working processes [6]. In AM process, including coating, through an incorporation of cold-working and its induced plastic deformation, undesired pores and/or micro-cracks can be closed while compressive residual stress can be built in the AM materials, so as to improve the mechanical performance and fatigue life [3].

2.2 Laser Treatment

Laser treatment delivers heat energy via photon-solid interactions directly on the surface of the workpiece. By changing the power density and pulse of the laser, the scanning rate, and the surface environments (e.g., in air or water), the photon-solid interactions can be tuned for laser cleaning (i.e., laser ablation), laser patterning/texturing, and laser melting [8]. Both the laser cleaning and the laser texturing have been applied to tune the surface energy for anti-fouling and/or anti-icing applications. Laser melting has recently been employed to treat the surface layer of metallic coatings produced by cold spray for corrosion and wear protections [9, 10].

3 Protective Coating

Coating is to deposit extrinsic materials on the surface of workpiece, e.g., thermal barrier on Ni-superalloy, for providing desired protections [11]. Both the coating material and the coating process have to be selected or designed based on the functional purpose.

3.1 Laser-Cladding

Laser-cladding is a direct energy deposition method that has been developed for welding and coating on metallic alloys [12]. Both powder and wire source materials can be used in this process. Recent studies show that carbon fibres can be mixed into the feedstock or separately fed in wire structure into the melting pool in laser-cladding process to produced novel coatings with improved performance and/or functions. Because laser melting of the source materials is direct on the surface of the substrate, metallic bonds are typically formed between the coating and the substrate. However, dilution of the substrate elements into the coating layer may occur and the rapid heating and cooling nature may cause cracks during the solidification of the melting pool. The process has to be optimized to minimize the dilution while the cracking has to be avoided for high performance coating.

3.2 Twin-Wire Arc Spray (TWAS)

Different from laser-cladding that melts the source materials direct on the surface of the substrate, TWAS melts the wire tips by electrical arc, which is generated via applying direct voltage on the wires with their tips being placed close to each other, at a certain distance away from the surface of the substrate. The melt droplets are blew by fast-flowing gas (i.e., atomization gas) towards the substrate surface to form the coating. Since the heat load carried by the droplets onto the substrate is not enough to melt the surface, substrate dilution is absent and the bonding between the substrate and the coating layer is mechanically rather than metallically, leading to a minor deformation of workpiece.

4 Conclusion and Outlook

In conclusion, we have briefly discussed surface enhancement and protective coating through the examples of RHP, SSP, laser-cladding, and TWAS. It is suggested that special care should be taken when looking at the repeatability and reliability due to the large number of processing parameters, especially those of the coating processes. Developments of new materials and compatible process integrations could have important consequence when innovating surface engineering for advanced industrial applications.

Acknowledgement. This work is partly supported by A*STAR RIE2020 advanced manufacturing and engineering (AME) programmatic grant through the structural metal alloys program (SMAP, Grant no. A18B1b0061).

References

1. Mosallanejad, M.H., Niroumand, B., Aversa, A., Saboori, A.: In-situ alloying in laser-based additive manufacturing processes: a critical review. J. Alloy. Comp. **872**, 159567 (2021)
2. Jahr, H., Li, Y., Zhou, J., Zadpoor, A.A., Schröder, K.-U.: Additively manufactured absorbable porous metal implants – processing. Alloy. Corros. Behav. Front. Mater. **8**(111), 628633 (2021)
3. Watanabe, H., Murata, T., Nakamura, S., Ikeo, N., Mukai, T., Tsuchiya, K.: Effect of cold-working on phase formation during heat treatment in CrMnFeCoNi system high-entropy alloys with Al addition. J. Alloy. Comp. **872,** 159668 (2021)
4. Behera, A., et al.: Effect of cold work on microstructure and corrosion properties of 304L In: Pant, P., Mishra, S.K., Mishra, P.C. (eds.) Advances in Mechanical Processing and Design. Lecture Notes in Mechanical Engineering, pp. 433–439. Springer, Singapore (2021). https://doi.org/10.1007/978-981-15-7779-6_37
5. Liu, H., et al.: Effects of robotic hammer peening on structural properties of Ni-based single-crystal superalloy: dislocation slip traces and crystallographic reorientations. Metall. Mater. Trans. A **51**(6), 3180–3193 (2020). https://doi.org/10.1007/s11661-020-05734-z
6. Liu, H., Ivan Tan, C.K., Wei, Y., Lim, G.W., Cheng, W.S., Maharjan, N.: Robotic hammer peening-induced martensite in austenitic steels: spatial distributions of plastic deformation and phase transformation. Procedia CIRP **87**, 297–301 (2020)
7. Liu, H., Wei, Y., Tan, C.K.I., Ardi, D.T., Tan, D.C.C., Lee, C.J.J.: XRD and EBSD studies of severe shot peening induced martensite transformation and grain refinements in austenitic stainless steel. Mater. Character. **168,** 110574 (2020)
8. Liu, H., Wei, Y., Ivan Tan, C.K., Lim, S.H., Zhang, L.: Laser-treatment-induced morphology and structure modifications of stainless steel: element segregations and phase evolutions. Mater. Chem. Phys. **266**, 124570 (2021)
9. Podrabinnik, P.A., Shishkovsky, I.V.: Laser post annealing of cold-sprayed Al–Ni composite coatings for green energy tasks. Procedia IUTAM **23**, 108–113 (2017)
10. Olakanmi, E.O., Doyoyo, M.: Laser-assisted cold-sprayed corrosion- and wear-resistant coatings: a review. J. Therm. Spray Technol. **23**(5), 765–785 (2014). https://doi.org/10.1007/s11666-014-0098-x
11. Liu, H., Chen, S., Tan, D.C.C., Lee, C.J.J., Yao, K.: Hot corrosion studies of a salt-coated Ni-based superalloy under flowing wet air and sulfur vapor ambient. Mater. Corros. **71**(10), 1608–1618 (2020)
12. Liu, H., Tan, C.K.I., Wei, Y., Lim, S.H., Lee, C.J.J.: Laser-cladding and interface evolutions of inconel 625 alloy on low alloy steel substrate upon heat and chemical treatments. Surf. Coat. Technol. **404,** 126607 (2020)

Development of Dry Lubricant Nickel-MoS_2 Composite Through Brush Plating

Yinsong Huang(✉), Zhaohong Huang, and Yujie Zhou

Singapore Institute of Manufacturing Technology, 2 Fusionopolis Way, #08-04, Innovis, Singapore 138634, Singapore
{huangys,zhhuang,yjzhou}@simtech.a-star.edu.sg

Abstract. Dry lubricant Nickel-MoS_2 composites were developed through brush plating by dispersion MoS_2 particles in the electrolyte. The good distribution of lubricant MoS_2 particles in the nickel matrix was confirmed by surface and cross-section analysis. The co-deposited MoS_2 can be up to 35 vol % in the coating with 45 g/L particle loading in the electrolyte, and the sliding coefficient of friction dramatically decreases from about 0.5 for nickel only to less 0.1 for Ni-MoS_2 composite while the counter material is bare stainless steel at 9.8 N loading force. The excellent dry lubrication property presents its application potential in preventing seizing and galling.

Keywords: Dry lubricant · Low friction · MoS_2 · Brush plating · Metal matrix composite

1 Introduction

Incorporating inert particles in a metallic matrix to improve the base metal coating performance are widely investigated in the past years [1–8].

In this study, we aimed to develop dry lubricate Ni-MoS_2 coatings through dispersion MoS_2 solid particles into brush nickel plating electrolytes with aid of selected surfactants. The Ni-MoS_2 co-deposition process, the effects of particle loading and its distribution on composite coatings are studied. Coating morphologies, hardness, compositions, coefficient of friction properties are evaluated.

2 Experimental

The MoS_2 powder (McLube MoS_2-100, 0.65–0.8 μ with a lamellar structure) used in this study is purchased from McGee Industries, Inc. The basic Nickel brush plating electrolyte is home formulated with a pH near neutral, which was prepared by dissolving reagent grade chemicals in deionized water. Triton X-100 wetting agent was selected to prevent agglomeration and keep uniform dispersion of the molybdenum disulfide particles in the bath, and the suspension was evaluated by ζ-potentials measurement. Ø50 × 5 mm mild steel plates were used as plating substrate; Engineering graphite

Y. Wei and S. Chng (Eds.): INCASE 2021, LNME, pp. 142–145, 2022.
https://doi.org/10.1007/978-981-16-5763-4_31

was used as brush anode material; 3M Scotchbrite was chose as anode wrap material to absorb and keep electrolyte during brush plating.

The coating hardness was measured using Matsuzawa MMT-X3 micro-hardness indentation. The coating phase and crystalline information was analyzed by Bruker D8 Discover X-ray diffraction (XRD. The morphology and cross-section of the coatings were observed using Carl Zeiss EVO50 scanning electron microscopy (SEM) and/or optical microscope. Coefficient of friction (COF) wear testing was carried out with CETR Universal Material Tester (UMT) Multi-functional Micro-Tribometer under unlubricated condition at room temperature in ambient air environment, and the stainless steel ball is used as counter material, the total testing wear distance is 120 m.

3 Results and Discussions

Figure 1 shows the ζ-potentials of 1 g/L MoS_2 particle dispersion with aid 25 mg/L Triton X-100 surfactant at different pH water solutions, it can be concluded that the selected MoS_2 particle can be well suspended (ζ-potentials are 40 mV) if the pH is above 4.0.

Figure 2 listed the HV100 measurement results of Ni-MoS_2 composite coatings brush plated with different particle loading electrolytes. The coating hardness is reduced with the amount of incorporated MoS_2 dry lubricant particle from pure Nickel 469 HV100 to–100 HV100 (for coating obtained with 45 g/L MoS_2 loading electrolyte).

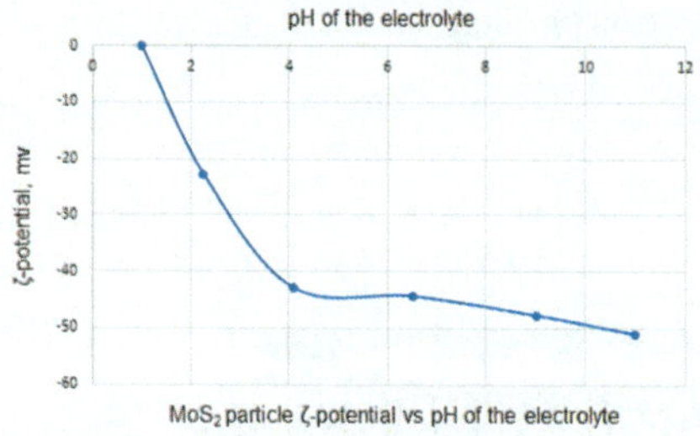

Fig. 1. pH effects on ζ-potentials of MoS_2 dispersion with aid of Triton X-100.

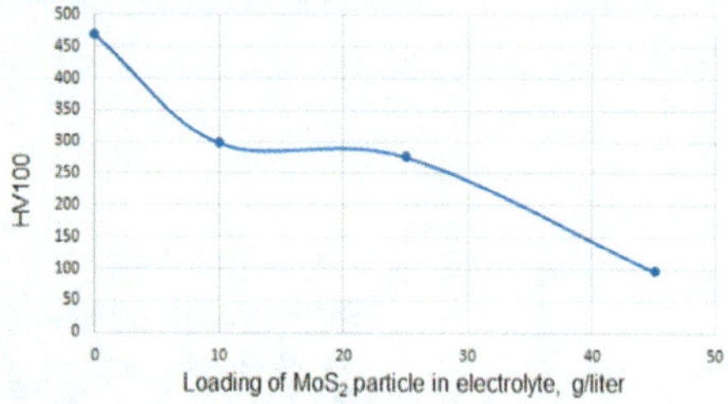

Fig. 2. Micro-hardness HV100 of Ni-MoS_2 composite coatings

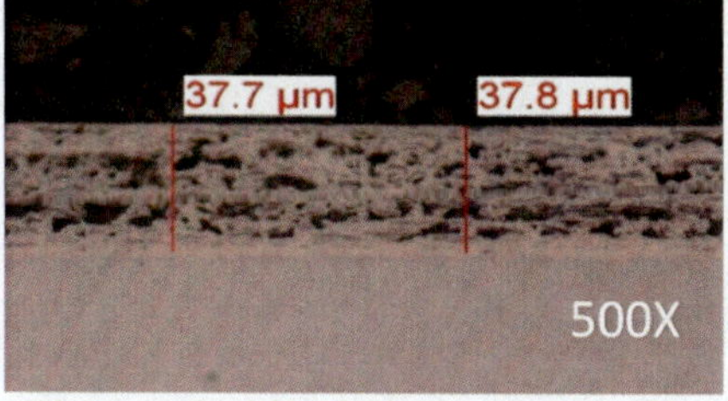

Fig. 3. Cross-section of brush plated Ni-MoS_2 composite

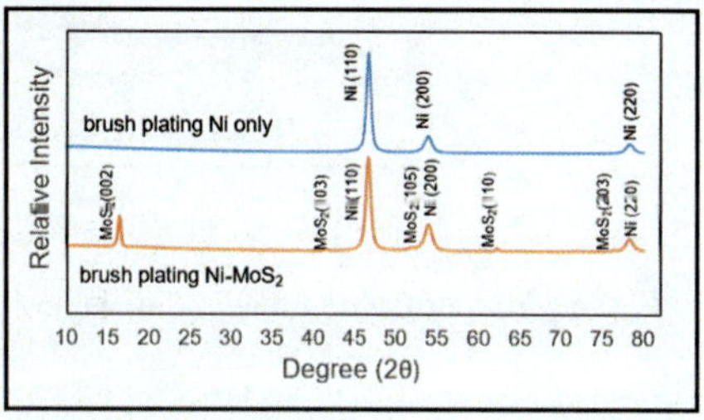

Fig. 4. XRD pattern of brush plated Ni and Ni-MoS_2 composites

Both cross-section observation (Fig. 3) and XRD pattern analysis (Fig. 4) are confirmed co-deposition of solid MoS_2 in the composite coatings. Sharp peaks of brush plated nickel coatings which correspond to Ni (100), Ni (200) and Ni (220) plane respectively. Other peaks of Ni-MoS_2 composite coating are resulted from the co-deposited MoS_2 particles, which correspond to the plane of MoS_2 planes of (002), (103), (105), (110) and (203) respectively [9]. In the meanwhile, the peaks correspond to nickel planes are hardly effected.

An EDX composition analysis (Fig. 5) shows that the MoS_2 is up to 22 wt %, corresponding to about 33–35 vol % (converted based on −5.0 g/cm^3 density of MoS_2).

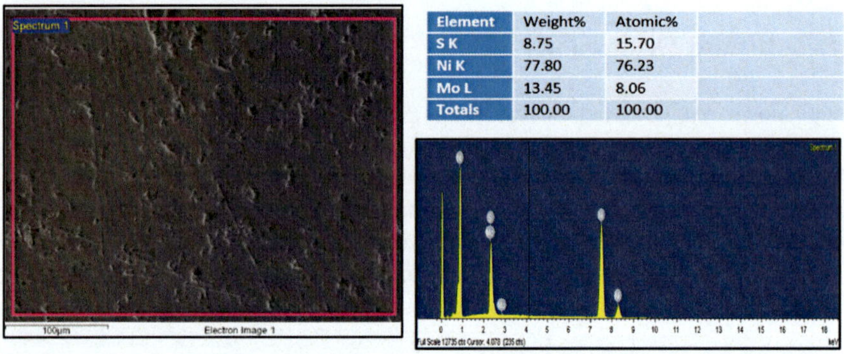

Element	Weight%	Atomic%
S K	8.75	15.70
Ni K	77.80	76.23
Mo L	13.45	8.06
Totals	100.00	100.00

Fig. 5. EDX analysis of Ni-MoS_2 composite (MoS_2 loading: 45 g/L in electrolyte)

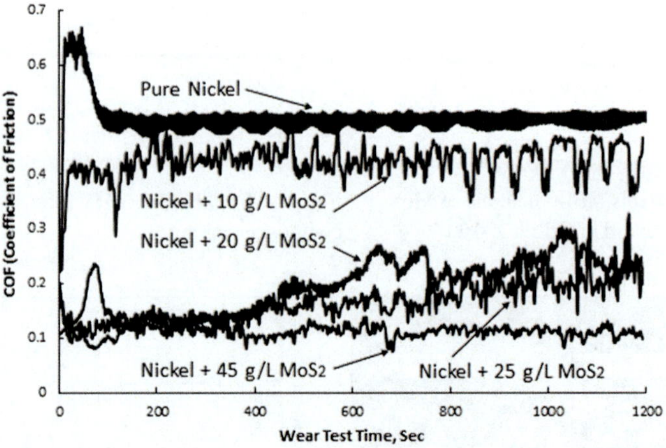

Fig. 6. Effect of MoS_2 loading in electrolyte Ni-MoS_2 composite coefficient of friction

Figure 6 shows the effect of MoS_2 loading in brush plating electrolyte on the coating lubricity. It can be seen that with increase of the MoS_2 particle loading in the brush plating electrolytes, the coefficient of friction (COF) of deposited coatings (measured at 9.8 N loading force with bare stainless steel as the counter material) decreases gradually, to even below 0.1. More importantly, when the MoS_2 loading is increased up to 45 g/L in the plating solution, the resulted Ni-MoS_2 composite coating shows very stable lubricity across the testing period. It was reported in the previous study on the MoS_2 composite coating [10] that, as the load increases, the friction coefficient of composite coating incorporated MoS_2 will further decrease due to the formation of a lubricating transfer layer on the counterpart surface during sliding. It is expected that the Ni-MoS_2 coatings by brush plating could reach much lower COF than 0.1 at higher load forces.

4 Conclusions

Ni-MoS_2 composites with MoS_2 content ranging from 0–35 vol% were successfully prepared by brush plating nickel through dispersion 0–45 g/L MoS_2 particles in a near neutral nickel electrolyte. The effects of MoS_2 content on the friction Ni-MoS_2 composites were evaluated. COF can be less than 0.1 while the counter material is stainless steel. The results demonstrated that the developed brush plating dry lubricant coatings can be an alternative to fluid lubricants for minimizing friction and preventing seizing and galling.

References

1. Grosjean, A., Rezrazi, M., Takadoum, J., Bercot, P.: Hardness, friction and wear characteristics of Nickel-SiC electroless composite deposits. Surf. Coat. Technol. **137**, 92–96 (2001)
2. Huang, W., Zhao, Y., Wang, X.: Preparing a high-particle-content Ni/diamond composite coating with strong abrasive ability. Surf. Coat. Technol. **235**, 489–494 (2013)
3. Karta, M., Aysal, M., Gul, H., Alp, A., Akbulut, H.: Effect of surfactant concentration in the electrolyte on the tribological properties of nickel-tungsten carbide composite coatings produced by pulse electro co-deposition. Appl. Surf. Sci. **354**, 328–336 (2015)
4. Wang, F., Arai, S., Endo, M.: Electrochemical preparation and characterization of nickel/ultra-dispersed PTFE composite films from aqueous solution. Mater. Trans. **45**(4), 1311–1316 (2004)
5. Plumiera, F., Chassaingb, E., Terwagnec, G., Delhallea, J., Mekhalif, Z.: Electrolytic co-deposition of a nickel/fluorographite composite layer on polycrystalline copper. Appl. Surf. Sci. **212–213**, 271–278 (2003)
6. Shi, L., Sun, C., Liu, W.: Electrodeposited nickel–cobalt composite coating containing MoS_2. Appl. Surf. Sci. **254**, 6880–6885 (2008)
7. Moonir-Vaghrfi, S.M., Saatchi, A.: The effect of agitation on electroless nickel-phosphorus-molybdenum disulfide composite plating. Metal Finish. **95**, 102–106 (1997)
8. Podgornik, B., Kosec, T., Kocijan, A., Donik, C.: Tribological behaviour and lubrication performance of hexagonal boron nitride (h-BN) as a replacement for graphite in aluminium forming. Tribol. Int. **81**, 267–275 (2015)
9. Zhou, N., Wang, S., Walsh, F.C.: Effective particle dispersion via high-shear mixing of the electrolyte for electroplating a nickel-molybdenum disulphide composite. Electrochim. Acta **283**, 568–577 (2018)
10. Fox, V., Hampshire, J., Teer, D.: MoS_2/metal composite coatings deposited by closed-field unbalanced magnetron sputtering: tribological properties and industrial uses. Surf. Coat. Technol. **112**, 118–122 (1999)

Effects of Spray Angle on Mass, Thickness, Porosity & Residual Stress in Cold-Sprayed Ti-6Al-4V Coatings

Zheng Zhang[1](✉), Debbie Hwee Leng Seng[1], Tzee Luai Meng[1], Siew Lang Teo[1], Zhi-Qian Zhang[2](✉), Boon Hee Tan[3], Qizhong Loi[3], and Jisheng Pan[1](✉)

[1] Institute of Materials Research and Engineering, A*STAR (Agency for Science, Technology and Research), #08-03, 2 Fusionopolis Way, Innovis, Singapore 138634, Singapore
{zhangz,debbie-seng,Meng_Tzee_Luai,sl-teo, js-pan}@imre.a-star.edu.sg

[2] Institute of High Performance Computing, A*STAR (Agency for Science, Technology and Research), #16-16, 1 Fusionopolis Way, Connexis, Singapore 138632, Singapore
zhangz@ihpc.a-star.edu.sg

[3] Advanced Remanufacturing and Technology Centre, A*STAR (Agency for Science, Technology and Research), CleanTech Two, #01/01, 3 Cleantech Loop, Singapore 637143, Singapore
{tanbh,loiq}@artc.a-star.edu.sg

Abstract. Dense and adherent Ti-6Al-4V (Ti64) coatings can be fabricated by cold spray (CS). To deposit Ti64 coatings on uneven surfaces of actual components via CS, the effect of different impact angles from 50 to 90° (with respect to the surface plane) has been investigated for deposition of spherical Ti64 powders on Al6061-T6 substrates using nitrogen as process gas. The development of the coating's mass, thickness, porosity and residual stress with impact angles have been analyzed.

Keywords: Cold spray · Ti-6Al-4V · Impact angle · Porosity · Residual stress

1 Introduction

Cold spray (CS) is one of the thermal spray techniques that utilizes the lowest process temperature (≤ 1100 °C) to soften the powders before impacting them onto the substrates. The high velocity impact induces shear instability between powder and substrate, which causes fragmentation of surface oxide and the formation of metallic bonding between the powder and underlying substrate [1]. A layer of powder coating can be formed or bulk components can be fabricated via repeated spray.

CS has been employed to deposit a variety of metallic powders (Al [2], Cu [3], Ti [4], Ni [5], etc.) as well as metallic-ceramic-composite powders (Al-Al_2O_3 [6]) onto components/substrates made from the same or similar materials. Among these alloys, Ti-6Al-4V or Ti64 alloy is an important alloy widely used in automotive [7], aerospace

Y. Wei and S. Chng (Eds.): INCASE 2021, LNME, pp. 146–149, 2022.
https://doi.org/10.1007/978-981-16-5763-4_32

[8] and medical [9] industries due to its high strength-to-weight ratio and excellent corrosion resistance. To repair Ti64-based high value components, CS could be a promising technique to restore near-net shape large dimensional components at ambient condition. Its low temperature process also avoids undesired powder oxidation and phase transformation during deposition.

To simulate the CS deposition of pure Ti powders on odd geometries, a few researchers have investigated the influence of impact angle [10, 11], gas temperature[11], substrate hardness [10] and gas type [12]. In this work, a detailed study of different spray angles from 50 to 90° when spraying Ti64 onto Al6061-T6 substrate was carried out to complement the existing knowledge on the effects of spray angle. The coating's mass, thickness, porosity and residual stress were investigated in order to understand the effect of the spray angle for better deployment of CS in actual repair applications.

2 Methods

Spherical Ti64 powders from Plasma Giken (D50 = 22.1 μm, part number: PG-AMP-1120) were sprayed onto Al6061-T6 substrate coupons ($5 \times 5 \times 1$ cm^3) via Plasma Giken PCS1000 cold spray system. N_2 pre-heated to 800 °C at 5.5 MPa was used as propellant gas. At a standoff distance of 3 cm, eight passes of Ti64 was deposited by traversing the nozzle head across the entire coupon at an estimated powder feed rate of 8 g/min to build up the coating. The nozzle was set at four angles of 50, 60, 70 and 90° to the surface plane.

After CS deposition, the coupons' surface morphology, weight gain, surface average roughness (R_a) and surface residual stress were measured via JEOL JSM-7600F field emission scanning electron microscope (SEM), weighing balance with an accuracy of 0.1 mg, KLA Tencor P-10 surface profilometer and Stresstech Xstress G3 X-ray stress measurement, respectively. The cross-section surfaces of the Ti64 coatings were mechanically polished using SiC sandpaper to grit size 4000, followed by MD Mol (3 μm) and MD Nap OPS (0.3 μm) final polishing. Images of the polished cross-section surfaces were taken by Olympus BX53M optical microscope. The coating thickness and porosity were subsequently measured using ImageJ software.

3 Results and Discussion

The surface morphologies of the four coatings are similar (Fig. 1(a–d)). Spherical powders are observed on the surface for all coatings, which is evidence of non-melting of the initial Ti64 powders even after passing through the pre-heated N_2 gas. However, the cross-sections of Ti64 coatings reveal interesting features related to spray angle (Fig. 1(a′–b′)).

The cross-sectional images show that the porosity decreased with increase in impact angle (14.7% @ 50° impact angle, Fig. 1(a′) to 1.8% @ 90° impact angle, Fig. 1(d′)). In addition, the pores at low impact angles (50 and 60°) are largely aligned in the direction of the impact angles (Fig. 1(a′ and b′)). The higher impact angles clearly densified the coating, reducing porosity (Fig. 2(a)). The decrease of porosity with increasing spray angle (Fig. 2(a)) has been similarly observed for deposition of pure Ti powders on $AlMg_3$

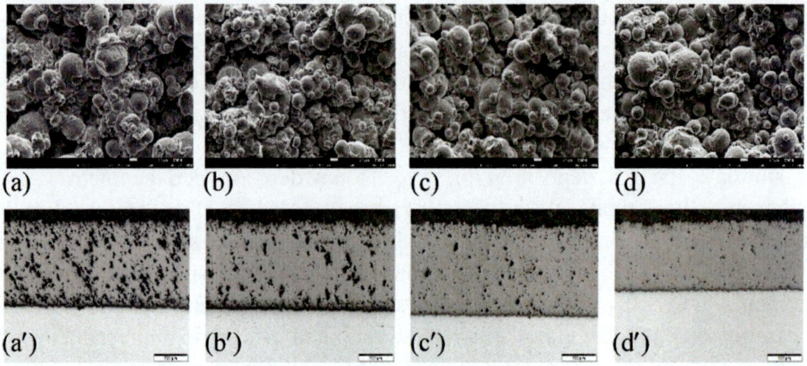

Fig. 1. (a–d) SEM images of top surface and (a′–d′) optical microscope images of cross-section surface for Ti64 coatings deposited at impact angles of (a & a′) 50, (b & b′) 60, (c & c′) 70 and (d & d′) 90°.

[11], Al, Cu and stainless steel [10] substrates. It shows that porosity is dominantly determined by spray angle and not much affected by substrate material.

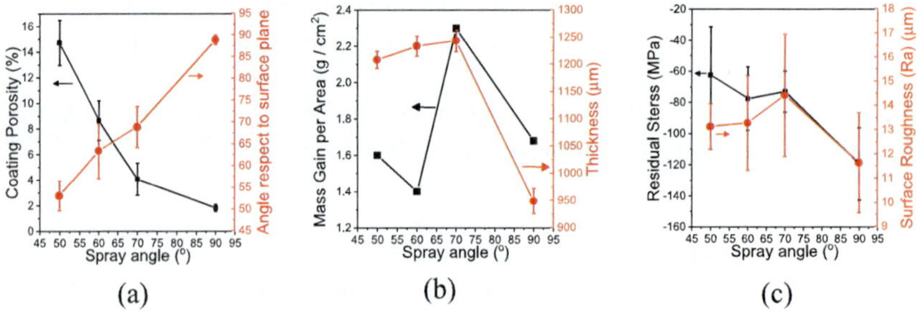

Fig. 2. (a) Coating porosity and alignment of the pores, (b) mass gain per area and coating thickness, (c) residual stress and surface roughness of Ti64 coatings with respect to spray angle.

The perpendicular impact at 90° spray angle led to the thinnest coating (Fig. 2(b)). This is in sharp contrast to the deposition of pure Ti powder on Cu substrate which led to the thickest coating with 90°spray angle [10]. This opposite behavior is still under investigation. With 90° impact angle, a significant peening effect was also induced, which led to a relatively smooth surface with the lowest surface roughness (11.6 μm) and the highest compressive residual stress (CRS, −119.3 MPa) among the four impact angles investigated (Fig. 2(c)). The impact angle of 70° resulted in the highest weight gain (Fig. 2(b)) and highest surface roughness (Fig. 2(c)). The surface CRS decreased with decreasing spray angle due to a lower vertical vector impact at a lower spray angle (Fig. 2(c)), which resulted in reduced densification of subsequent coating and unflattened powder morphology.

4 Conclusion

Spray angle has a significant effect on the porosity of cold-sprayed coatings. A decrease in spray angle will considerably increase the porosity in the Ti64 coatings while reducing the surface CRS. At spray angles of 60° and below, pores largely aligned in the direction of the spray angle are formed inside the coating. Such porosity can deteriorate the coating's adhesion strength and wear resistance tremendously. Therefore, to achieve dense and continuous Ti64 coatings during the actual repairing process of components, it would be very critical to control the impacting angle of the CS gun, with impact angles higher than 70° recommended.

Acknowledgements. This research is supported by the Agency for Science, Technology and Research (A*STAR) under Machining Learning Assisted Control of Metal Cold Spray and Shot Peening Processes (A1894a0032), Structural Metal Alloy Program (A18b1B0061) and Accelerated Materials Development for Manufacturing (A1898b0043).

References

1. Schmidt, T., et al.: From particle acceleration to impact and bonding in cold spraying. J. Therm. Spray Technol. **18**, 794–808 (2009)
2. Rokni, M.R., Widener, C.A., Champagne, V.R.: Microstructural evolution of 6061 aluminum gas-atomized powder and high-pressure cold-sprayed deposition. J. Therm. Spray Technol. **23**(3), 514–524 (2013). https://doi.org/10.1007/s11666-013-0049-y
3. Koivuluoto, H., Coleman, A., Murray, K., Kearns, M., Vuoristo, P.: High pressure cold sprayed (HPCS) and low pressure cold sprayed (LPCS) coatings prepared from OFHC cu feedstock: overview from powder characteristics to coating properties. J. Therm. Spray Technol. **21**, 1065–1075 (2012)
4. Li, W.Y., et al.: Ti and Ti-6Al-4V coatings by cold spraying and microstructure modification by heat treatment. Adv. Eng. Mater. **9**, 418–423 (2007)
5. Zhang, Z., et al.: Cold spray deposition of inconel 718 in comparison with atmospheric plasma spray deposition. Appl. Surf. Sci. **535**, 147704 (2021)
6. Lee, H.Y., Jung, S.H., Lee, S.Y., You, Y.H., Ko, K.H.: Correlation between Al_2O_3 particles and interface of Al-Al_2O_3 coatings by cold spray. Appl. Surf. Sci. **252**, 1891–1898 (2005)
7. Faller, K., Froes, F.H.: The use of titanium in family automobiles: current trends. JOM **53**, 27–28 (2001)
8. Boyer, R.R.: An overview on the use of titanium in the aerospace industry, Mater. Sci. Eng. A **213**, 103–114 (1996)
9. Rack, H.J. Qazi, J.I.: Titanium alloys for biomedical applications, Mater. Sci. Eng. C **26** 1269–1277 (2006)
10. Yin, S., Suo, X., Su, J., Guo, Z., Liao, H., Wang, X.: Effects of substrate hardness and spray angle on the deposition behavior of cold-sprayed Ti particles. J. Therm. Spray Technol. **23**(1–2), 76–83 (2013). https://doi.org/10.1007/s11666-013-0039-0
11. Binder, K., Gottschalk, J., Kollenda, M., Gärtner, F., Klassen, T.: Influence of impact angle and gas temperature on mechanical properties of titanium cold spray deposits. J. Therm. Spray Technol. **20**, 234–242 (2011)
12. Wong, W., Irissou, E., Ryabinin, A.N., Legoux, J.-G., Yue, S.: Influence of helium and nitrogen gases on the properties of cold gas dynamic sprayed pure titanium coatings. J. Therm. Spray Technol. **20**, 213–226 (2011)

Corrosion and Mechanical Behavior of Plasma Electrolytic Oxidation Coated Al Alloy

Junyi Liu(✉), Xiaohu Huang, Tzee Luai Meng, and Shijie Wang

Institute of Materials Research and Engineering, A*STAR (Agency for Science, Technology and Research), #08-03, 2 Fusionopolis Way, Innovis, Singapore 138634, Singapore
{liu_junyi,huangxh,Meng_Tzee_Luai,sj-wang}@imre.a-star.edu.sg

Abstract. Alumina coating was deposited as a protective layer on Al alloy by plasma electrolytic oxidation (PEO) process under unconventionally high applied voltage and high $SiO_3{}^{2-}$ concentration. Scanning electron microscopy (SEM) with energy dispersive X-ray analysis (EDX) was used to analyze the microstructure and thickness of the coating. Corrosion, hardness and adhesion properties linked to its protective performance were studied by linear polarization, nanoindentation and ASTM D3359 adhesion test. The high voltage contributed to the formation of ceramic coating consisting of α-phase Al_2O_3 and mullite. Due to the compact crystalized structure produced by PEO, the corrosion resistance was significantly improved. The proposed coating also exhibited high hardness and excellent adhesive strength.

Keywords: Protective coating · PEO · Anodization · Al alloy

1 Introduction

7-series Al alloy (AA 7xxx) has been extensively used in aerospace industry due to its high strength provided by the added strengthening phases. Major alloying elements and impurities include Zn, Mg, Cu and Fe [1, 2], which form several intermetallics, determining the mechanical properties of the alloy. However, the intermetallics are also the initial sites leading to localized corrosion, which may seriously reduce the strength of Al during the service period. Therefore, suitable surface treatments are necessary to improve their corrosion and wear resistance.

Currently the most popular surface treatments to increase the corrosion and wear resistance include chemical conversion, anodization, electrochemical deposition and sol–gel process [3]. Among them, anodization producing a thin oxidized layer on the matrix, are extensively used in the aerospace industry in the past few decades. Plasma electrolytic oxidation (PEO) process is originated from traditional anodization. In this process, plasma discharge occurs at high applied voltages which leads to a more compact and harder oxides than other common surface treatments. Although the properties of many PEO coatings on light alloys has been reported in the literature, report about the effect on corrosion and mechanical behaviors by super high voltage (above 600 V) PEO coating is quite limited. Unconventionally high applied voltage provides violent plasma

Y. Wei and S. Chng (Eds.): INCASE 2021, LNME, pp. 150–154, 2022.
https://doi.org/10.1007/978-981-16-5763-4_33

and extremely high reaction temperature, which may cause phase change of alumina and thus provide the ceramic coating with new properties [4]. Beside applied voltage, electrolyte composition is another important parameters affecting microstructure and properties of the coating. Alkali electrolyte containing SiO_3^{2-} ions was used in many research works on PEO coating. The concentration of SiO_3^{2-} in the reported literature is mostly in the range of 1 g/L to 6 g/L. Yang *et al.* and Becerik *et al.* investigated the effect of SiO_3^{2-} concentration in the electrolyte and concluded that increase in the concentration improves the corrosion resistance of PEO coating [5, 6]. However, the effect on the mechanical properties of the coating was not reported in their study.

In this study, PEO coating was fabricated on AA 7075 using unconventionally high voltage and high SiO_3^{2-} concentration electrolyte to explore their combined surface enhancement effect to corrosion and mechanical properties. These properties are significantly important for the PEO coating acting as a protective layer for various applications in automotive, aerospace and marine industries.

2 Methods

The AA 7075 T6 coupons with 2 mm thickness were used in this study. PEO process was carried out using a AC high voltage power supply. Al alloy specimens were used as anode, and one stainless steel 316 panel was positioned as cathode. In the process, applied voltage was 700 V (starting current of 0.15 A/cm^2) with 50 Hz frequency for 20 min. The 1 L electrolyte contains 2 g/L KOH and 10 g/L Na_2SiO_3. The conventional anodization was carried out under 15 V in 5 g/L H_2SO_4 (200 mL) for 20 min.

X-ray diffraction (XRD) was performed using a General Area Detector Diffraction with a Cu X-ray source while the JEOL Nova Nanosem 230 was used for the SEM analysis, with EDX. Hardness and Young's modulus of the samples' cross-section were evaluated by Agilent G200 Nano-indentation using a continuous stiffness measurement technique. The cross section of the samples was the polished prior to the test.

For the corrosion test, the AA 7075 sample was sealed in a sample holder with 1 cm^2 exposed area. Measurements were carried out with a three-electrode cell, comprising a working electrode (AA 7075), a counter electrode (Pt), and a reference electrode (Ag/AgCl) in a 3.5 wt.% sodium chloride (NaCl) solution. Applied potential is ±0.6 V with respect to open circuit potential (OCP), with a scan rate of 100 mV/min. The fitting was done in the polarization range ±10 mV versus corrosion potential.

3 Results and Discussion

The surface morphology of the PEO coating under 700 V applied voltage and 10 g/L SiO_3^{2-} concentration is shown in Fig. 1(a). The rough surface observed contains many grains and residual pores. These pores were generated due to the plasma spark channels. When the melt oxidization products erupted from the discharge channel, it quenched and deposited to form the porous rough surface. The XRD results shown in Fig. 1(b) indicated that the coating was crystallized and mainly contains α-phase Al_2O_3 and mullite ($3Al_2O_3{\cdot}2SiO_2$ or $2Al_2O_3{\cdot}SiO_2$). Due to the high temperature generated during PEO process, the molten alumina can react with silicate ions presented in the electrolyte and form a stable mullite phase [7], which has a more compact structure than alumina [8]. The coating thickness was around 10 μm as shown in the cross sectional image of the PEO on AA 7075 substrate in Fig. 2. The composition of the coating was also examined using EDX, proving the incorporation of Si in the ceramic alumina coating.

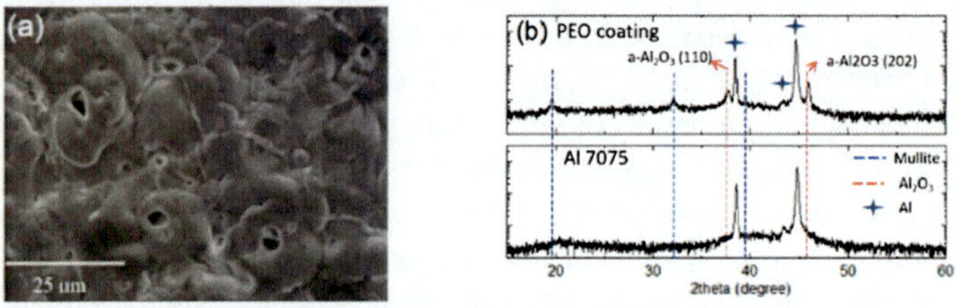

Fig. 1. (a) SEM images showing the surface morphology of PEO coating and (b) XRD analysis results of the PEO coating and bare AA7075

To illustrate the corrosion behaviors, the linear polarization curves obtained for the bare and PEO coated Al alloy in 3.5 wt % NaCl solution (Fig. 3). The bare AA7075 alloy had a low corrosion potential (Ecorr = −0.62 V) and a rather high corrosion current density (icorr = 55.2 μA/cm^2). After PEO treatment, the corrosion potential of the coated sample was increased to −0.3 V and the corrosion current density was significantly reduced to 2.3 μA/cm^2. The results indicated a reduction in corrosion current density by more than 20 times, suggesting that PEO coating can act as a good barrier offering enhanced corrosion protection for the base alloy.

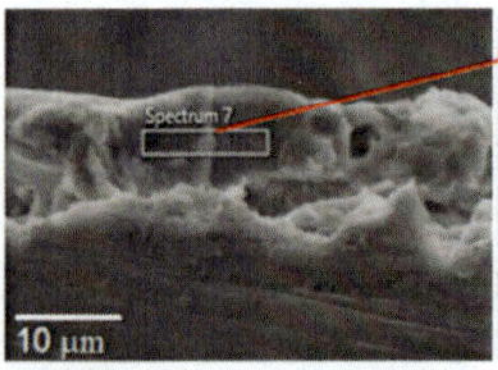

Element wt%	PEO	AA 7075
Al	39.9	80.8
O	47.4	1.7
Si	0.5	0
C	11.1	7.5
Others	1.1	Zn 6.2 Mg 2.0 Cu 1.8

Fig. 2. Cross section SEM images of PEO on Al alloy with the relative element composition by EDX

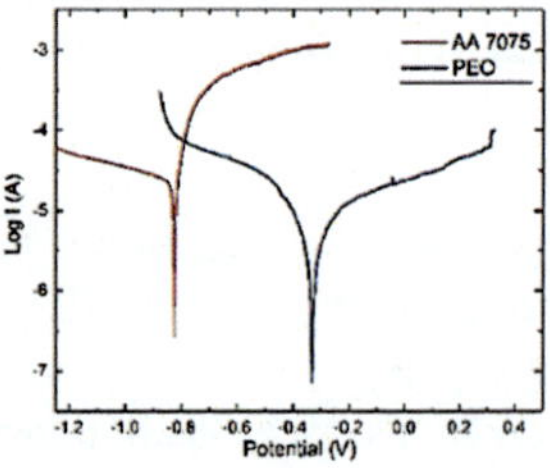

Fig. 3. Linear polarization plots for bare and PEO coated AA7075

Typical load-displacement curve of the AA7075 with/without PEO coating obtained from Nano-indentation measurements is shown in Fig. 4. The indentation hardness and Young's modulus values obtained for AA7075, PEO and conventional anodized coating were summarized in Table 1. The hardness of the PEO coating was improved by 3 and 6 times compared to that of AA7075 and conventional anodized coating, respectively. The modules of the PEO coating was improved as well. The results suggested that PEO coating could effectively improve the wear resistance of AA 7075.

Since mechanical properties of the PEO coating also depend on the adhesion between the coating and the substrate, adhesion test was performed on the coated samples. The

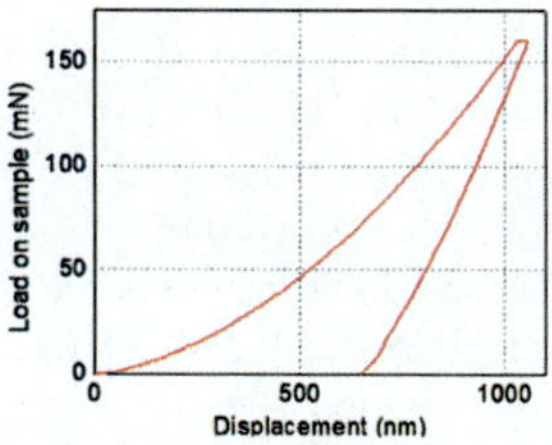

Fig. 4. Typical indentation curve of PEO coating sample

Table 1. Hardness and modulus data of Al matrix, PEO and conventional anodized coating

Sample	Test	Mean (GPa)
PEO	Hardness	7.90 ± 0.81
	Modulus	92.9 ± 9.52
Al 7075	Hardness	2.3 ± 0.05
	Modulus	84.5 ± 1.67
Conventional anodization with hot water sealing	Hardness	0.65 ± 0.10
	Modulus	17.1 ± 3.09

coating was tested with cutter blade and ASTM D3359 adhesive tape. The adhesion between the coating and Al coupon corresponded to a 5B grade (the highest) according to ASTM classification. There is no crack or chip of the coating after the quenching tests.

4 Conclusion

AA 7075 was deposited with ceramic alumina coating by PEO method under unconventionally high applied voltage and high SiO_3^{2-} concentration. The corrosion test shows that the corrosion resistance of AA7075 could be significantly improved by the proposed PEO treatment, and this is attributed to the compact crystalized structure produced. The coating exhibited high hardness and modulus compared to the base AA 7075 and conventional anodized coating. The 5B grade adhesion value obtained also indicates the coating exhibited excellent adhesion strength with the substrate.

References

1. Birbilis, N., Cavanaugh, M.K., Buchheit, R.G.: Electrochemical behavior and localized corrosion associated with Al7Cu2Fe particles in aluminum alloy 7075–T651. Corros. Sci. **48**(12), 4202–4215 (2006)
2. Venugopal, A., Srinath, J., Rama Krishna, L., Ramesh Narayanan, P., Sharma, S.C., Venkitakrishnan, P.V.: Corrosion and nanomechanical behaviors of plasma electrolytic oxidation coated AA7020-T6 aluminum alloy. Mater. Sci. Eng. A **660**, 39–46 (2016)
3. Huang, Y., et al.: Evaluation of the corrosion resistance of anodized aluminum 6061 using electrochemical impedance spectroscopy (EIS). Corros. Sci. **50**(12), 3569–3575 (2008)
4. Dehnavi, V., Liu, X.Y., Luan, B.L., Shoesmith, D.W., Rohani, S.: Phase transformation in plasma electrolytic oxidation coatings on 6061 aluminum alloy. Surf. Coat. Technol. **251**, 106–114 (2014)
5. Becerik, D.A., Ayday, A., Kumruoğlu, L.C., Kurnaz, S.C., Özel, A.: The effects of Na_2SiO_3 concentration on the properties of plasma electrolytic oxidation coatings on 6060 aluminum alloy. J. Mater. Eng. Perform. **21**(7), 1426–1430 (2012)

6. Yang, Y., Zhou, L.: Improving corrosion resistance of friction stir welding joint of 7075 aluminum alloy by micro-arc oxidation. J. Mater. Sci. Technol. **30**(12), 1251–1254 (2014)
7. Raj, V., Mubarak Ali, M.: Formation of ceramic alumina nanocomposite coatings on Al for enhanced corrosion resistance. J. Mater. Process. Technol. **209**(12), 5341–5352 (2009)
8. Liu, C., et al.: The correlation between the coating structure and the corrosion behavior of the plasma electrolytic oxidation coating on aluminum. Surf. Coat. Technol. **286**, 223–230 (2016)

Effectiveness of Cold Spray Coating on Cast Iron to Prevent Oxidation

Nataniel Yong Syn Tham[1,4], Yao Shian Chua[2], Kaiqiang Wu[1,2], Adrian Wei-Yee Tan[1,2,5], Sung Chyn Tan[3], and Wei Zhou[1,2(✉)]

[1] Rolls-Royce@NTU Corporate Laboratory, Nanyang Technological University, 65 Nanyang Drive, Singapore 63746, Singapore
wzhou@cantab.net, mwzhou@ntu.edu.sg
[2] School of Mechanical and Aerospace Engineering, Nanyang Technological University, 50 Nanyang Avenue, Singapore 639798, Singapore
[3] Rolls-Royce Singapore Pte. Ltd., 1 Seletar Aerospace Crescent, Singapore 797575, Singapore
[4] Advanced Remanufacturing and Technology Center (ARTC), 3 Cleantech Loop, Singapore 637143, Singapore
[5] Current Address: School of Engineering, University of Southampton Malaysia, 79200 Iskandar Puteri, Johor, Malaysia

Abstract. Cold spray is an emerging additive manufacturing process allowing particles to be coated onto surface of base material without melting. In this study, Diamalloy 1005 was successfully coated onto GJL-250 cast iron using a cold spray process, and extensive experiments were carried out to observe the effects of diffusion after heat treatment at 400, 750 and 850 °C for 3 h and to evaluate effectiveness of the coating in preventing oxide growth. It is found that the diffusion layer increases in thickness with increasing heat treatment temperature, and the diffusion layer thickness is 45 and 75 µm after the heat treatment at 400 and 850 °C respectively. The diffusion layer does not affect the effectiveness of the coating acting as a protective layer. This finding demonstrates the potential application of using cold spray coating as a protective layer for cast iron.

Keywords: Cold spray · Cast iron · Heat treatment · Oxide growth · Diffusion layer

1 Introduction

Cold spray is an additive manufacturing process which accelerates particles onto the surface to be coated, allowing the particles to adhere onto that surface and form a coating. GJL-250 is one of the most common types of cast iron desired for its stiffness, making it commonly used in engine blocks. However, it is often subjected to oxidation, leading to detrimental properties over time and subsequent replacement of the entire part.

Repairing cast iron has traditionally been a challenge because the high carbon content makes it crack easily. Diamalloy 1005, similar in composition to Inconel 625 is corrosion resistant and is often used to protect components from oxidation, increasing the life span of the part. However, little research has been conducted on the possibility of cold spraying

Y. Wei and S. Chng (Eds.): INCASE 2021, LNME, pp. 155–158, 2022.
https://doi.org/10.1007/978-981-16-5763-4_34

Inconel 625 on GJL-250 cast iron and the possible interactions between them at high temperatures. This is due to the current challenge in creating a coating from a relatively hard metal such as Inconel onto another hard surface. Adiabatic shear instability is often cited to be the mechanism of adhesion in cold spray, allowing jetting to form between the particles and surface and provide good mechanical interlocking. However, for powders and surfaces that are hard, this phenomenon does not always occur, as the particles would simply fall off the surface after impact. As such, this paper seeks to study the feasibility of coating Inconel 625 onto GJL-250 cast iron, followed by studying the diffusion boundary between at the interface of the two metals after heat treatment at different temperatures and evaluating their effectiveness in reducing corrosion.

2 Methods

2.1 Cold Spray Process

The GJL-250 substrate was prepared by grinding the surface before spraying to remove any oxidation layer. The feedstock used was Diamalloy 1005 powder from Oerlikon Metco with a mean size of 45 μm. A cold spray system (Plasma Giken, PCS 100) was used to achieve a coating thickness of 1.2 mm. The nozzle standoff distance was 35 mm and the nozzle traverse speed was 500 mm/s. Thereafter, a furnace (Lenton Furnace, UK) was used to carry out heat treatment at 400, 750 and 850 °C for 3 h before cooling to room temperature.

2.2 Materials Characterization

The as-sprayed and heat-treated samples were hot mounted and then ground and polished for further analyses. Subsequently, the cross-sections were observed under the optical microscope (Zeiss, Axioskop 2 Mat) and scanning electron microscope (JEOL, JSM-5600LV). Energy dispersive X-ray spectroscopy (EDX) was performed to determine elemental composition around the diffusion layer.

3 Results and Discussion

Figure 1 shows the EDX-mapping of the as-sprayed sample and heat-treated samples. It can be seen from the figure that higher heat treatment temperatures provided a greater diffusion layer between the interface of the substrate and coating, with the thickness being 45 μm and 75 μm for the 400 °C and 850 °C heat treatment. Occurrence of diffusion in the as-sprayed condition (30 μm in thickness) is attributed to the high substrate temperature during the spraying process.

Solid-state diffusion is based on Fick's second law which predicts how concentration changes with time. Increment of temperature imparts higher amounts of energy to allow exchange of atoms between metals due to their atomic vacancies, explaining the changes in concentration of elements across temperature. Furthermore, the rate of diffusion is different for various elements, with chromium diffusing at the fastest rate (0.343 ×

10^{-11} cm^2/s at 1050 °C) and molybdenum the slowest rate (0.144×10^{-11} cm^2/s at 1050 °C), according to a study on the activation energy and diffusion coefficient of various elements in Inconel 718 [1].

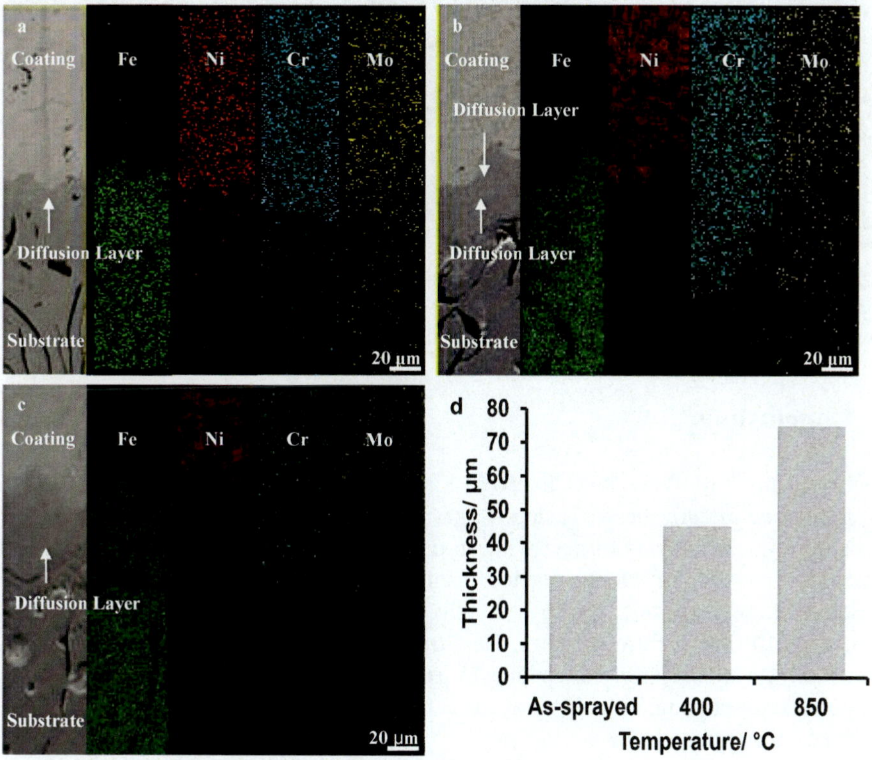

Fig. 1. EDX mapping of samples subjected to heat treatment at (a) As-sprayed; (b) 400 °C; (c) 850 °C; and (d) Comparison of thickness of diffusion layer.

In Fig. 2(a), it is apparent that there is no delamination at the interface between the cold sprayed coating and the cast iron, with a clear indication of Inconel 625 forming a jetting effect on the softer iron surface seen from the uneven interface layer. Faint outlines of inter-particle bonding were also observed, confirming good adhesion between the particles as they are deformed rather than spherical. Although a diffusion layer is present, Figs. 2(a) and (b) highlight the effect of a coating on the substrate, with both samples subjected to heat treatment at 750 °C for 3 h.

Pores in the coating amounted to 0.782%, which could allow oxygen to seep through to the cast iron. The absence of oxide shows that the coating is effective in acting as a protective layer, while oxide growth of approximately 100 µm is observed when there is no coating on cast iron. Oxide is formed due to the reaction of iron and oxygen and is a form of corrosion of cast iron, highlighting the advantage of having a coating to prevent it.

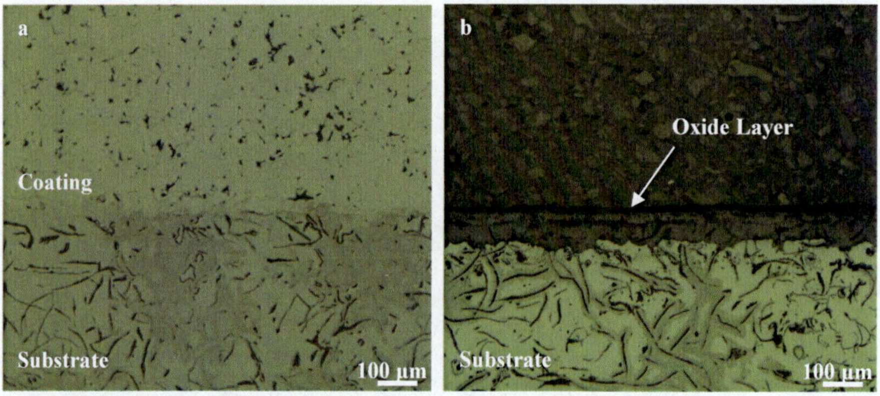

Fig. 2. Cross-section images taken under the optical microscope comparing oxide growth after heat treatment at 750 °C for 3 h (a) No observable oxide growth in the coated cast iron; and (b) Presence of oxide layer on the surface of uncoated cast iron.

4 Conclusions

Coating Diamalloy 1005 onto GJL-250 cast iron using cold spray process highlights the feasibility and effectiveness of using it as a protective layer against oxidation. Firstly, a coating of 1.2 mm was formed with no delamination on the interface between the coating and cast iron. It is found that increasing the heat treatment temperature increases thickness of the diffusion layer formed between the substrate and coating, with various elements diffusing at different rates. Heat treatment at 400 °C and 850 °C yielded a diffusion layer thickness of 45 µm and 75 µm. Despite presence of the diffusion layer and pores in the coating, the cold-sprayed coating is found to be able to prevent oxidation of the cast iron effectively.

Acknowledgements. The research was funded by National Research Foundation of Singapore, Rolls-Royce and Nanyang Technological University through research grant ARMS 1.1 (Advanced Metallization Coatings using Cold Spray).

Reference

1. Sohrabi, M.J., Mirzadeh, H.: Interdiffusion coefficients of alloying elements in a typical Ni-based superalloy. Vacuum **169**, 108875 (2019)

Cold Spray Coatings for Surface Enhancement of Inconel 718

Kaiqiang Wu[1,2], Sin Wei Chee[2], Wen Sun[1,2], Adrian Wei-Yee Tan[1,2,4], Sung Chyn Tan[3], and Wei Zhou[1,2(✉)]

[1] Rolls-Royce@NTU Corporate Laboratory, Nanyang Technological University, 65 Nanyang Drive, Singapore 637460, Singapore
kaiqiang001@e.ntu.edu.sg, wzhou@cantab.net, mwzhou@ntu.edu.sg
[2] School of Mechanical and Aerospace Engineering, Nanyang Technological University, 50 Nanyang Avenue, Singapore 639798, Singapore
[3] Rolls-Royce Singapore Pte. Ltd., 1 Seletar Aerospace Crescent, Singapore 797575, Singapore
[4] Current Address: School of Engineering, University of Southampton Malaysia, 79200 Iskandar Puteri, Johor, Malaysia

Abstract. Cold spray is an emerging additive manufacturing technology that is promising to be used for surface repair and enhancement. In this study, Inconel 713 coating was successfully deposited on Inconel 718 substrate using a high-pressure cold spray process. The microstructure and hardness of the coating were studied. The cold sprayed Inconel 713 coating had a very low porosity level due to the severe plastic deformation of the splats. Hardness of the coating is much higher than that of Inconel 718 substrate. The regions closer to the substrate-coating interface in the coating showed higher hardness value compared to those located further away from the interface. This phenomenon can be attributed to the peening effect during high-velocity impact of Inconel 713 particles with the Inconel 718 substrate. This study demonstrates that cold sprayed coatings have the potential to increase surface hardness and further improve wear resistance of the Inconel 718 substrate.

Keywords: Cold spray · Hardness · Inconel 718 · Inconel 713 · Surface enhancement

1 Introduction

Cold spray is an emerging technology that deposits particles onto substrate in the solid-state. Particles are accelerated to supersonic speeds by high pressure and high temperature gas via a convergent-divergent nozzle. Upon impact of particles on a substrate surface, thermal softening effect outweighs strain-hardening effect due to transformation of kinetic energy to thermal energy. Thus, both the particles and the substrate go through drastic plastic deformation and metallurgical bonding could form in the deformed regions.

Unlike thermal spray, the particle temperature in cold spray process remains much lower than the melting temperature, so there is little oxidation or phase transformation

Y. Wei and S. Chng (Eds.): INCASE 2021, LNME, pp. 159–162, 2022.
https://doi.org/10.1007/978-981-16-5763-4_35

in either coating or substrate, which makes cold spray promising for repair and additive manufacturing in various industries.

Inconel 718 (IN718) is the most widely used nickel-based superalloy due to its high strength and good corrosion resistance. However, parts made of IN718 can be degraded due to wear and corrosion. Inconel 713 (IN713) is a nickel-based alloy which possesses better mechanical properties and corrosion resistance at high temperatures compared with IN718. In this study, cold spray was used to deposit IN713 onto IN718 substrate to improve the surface hardness, while ductility of the substrate material remained the same.

2 Methods

2.1 Cold Spray Process

Gas atomized IN713 powder with the particle size ranging from 15 μm to 45 μm (Sandvik, Sweden) was used as the feedstock in the cold spray deposition process. Substrate used was IN718 alloy (Top Alloys, Singapore). Before the cold spray deposition, the substrate surfaces were ground with sandpaper and then cleaned in ethanol.

A high-pressure cold spray system (Impact Innovations GmbH, Germany) was used to do the deposition. Nitrogen (N_2) was used as the propellant gas which was preheated to 1000 °C and pressurized to 4.5 MPa. The De Laval nozzle of the system was held perpendicular to the substrate surfaces with a standoff distance of 30 mm from the nozzle exit. The nozzle transverse speed was 500 mm/s and the feed rate of the IN713 powder was around 48 g/min. Coating of a thickness around 300 μm was achieved with 10 layers of deposition.

2.2 Characterization

The coated samples were cut vertically to the coating-substrate interface. Subsequently, they were mounted in epoxy and then ground and polished. Final polishing was conducted using a colloidal silica suspension (~0.25 μm particle size). Optical microscopy (Zeiss Axioskop 2, USA) images of the IN713 coating were taken and then investigated with the ImageJ software for porosity calculation in compliance with the standard ASTM E2109-01 [1]. Backscattered electron (BSE) mode images were taken under the field emission scanning electron microscope (FE-SEM, Joel JSM-7600F, Japan) with a BSE detector.

Microhardness of the coating and substrate was measured using a Vickers hardness tester (Future-tech FM-300e, Japan) with a load of 500 g. For the hardness of coating, indentations were implemented 50 μm and 250 μm away from the coating-substrate interface. The hardness of the substrate was also measured in the same way.

3 Results and Discussion

Figure 1(a) is the optical image showing cross-section of the cold sprayed IN713 on IN718. A dense coating microstructure is seen with a very low porosity of 1.23%.

Because adiabatic shear instability (ASI) is triggered in the cold spray process, particles undergo severe plastic deformation upon impact. Jetting and splats formed during deformation tend to fill the voids and vacancies in the coating.

The interface between the coating and the substrate is undulating, indicating metallurgical bonding at the interface. Figure 1(b) is the BSE-SEM image taken at the interface with higher magnification. The dendrite structure of the splats in the coatings is heavily distorted, which is also the evidence of severe plastic deformation. In the substrate, microstructure in the region closer to the interface is distorted as well compared to the region away from the interface, which can be attributed to the peening effect when IN713 particles impacted the IN718 substrate.

Fig. 1. (a) Optical image of IN713 coating on IN718 substrate; (b) BSE-SEM image at a higher magnification showing the coating-substrate interface.

Figure 2 shows the hardness of the IN713 coating and IN718 substrate at various locations. Hardness of the coating is much higher than that of the substrate due to use of the harder coating material and perhaps partially due to strain accumulation and smaller grains in the coating [2]. It can be seen from the error bar that fluctuation of the hardness values of the coating is larger than that of the substrate. Defects in the coating may contribute to this scatter. In the microhardness tests, the indenter can interact with interparticle boundaries of poor cohesive strength and pores which are randomly distributed in the coating.

It is interesting to note that hardness values are higher when the hardness tests were done at a shorter distance of 50 μm away from the interface compared with those at 250 μm. This may be attributed to peening effect. In the substrate, regions close to the surface were shot-peened by the high-velocity particles, thus hardness in those regions increased. The regions further away from the surface were little affected. In the coatings, particles from subsequent layer impacted on previous layer also resulted in peening effect and made the previous layer harder. Since hardness is an indicator of wear resistance to some degree [3], therefore cold sprayed IN713 coatings is promising to improve wear resistance of IN718 substrates.

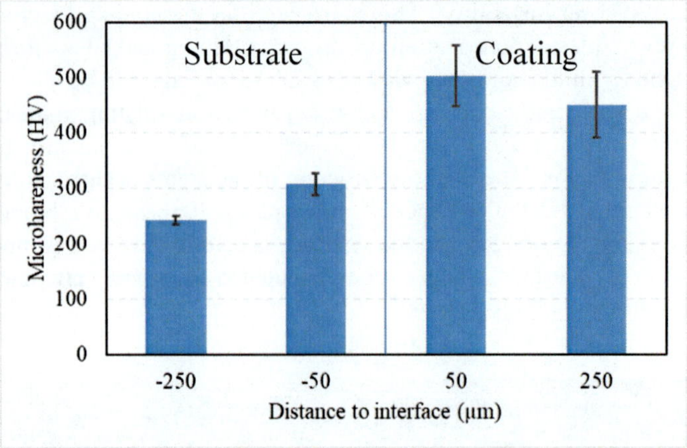

Fig. 2. Microhardness of IN713 coating and IN718 substrate at different locations.

4 Conclusions

In this study, IN713 coating was successfully deposited on IN718 substrate using a high-pressure cold spray system. Microstructural analysis demonstrated the high-quality cold-sprayed coating with low porosity level of 1.23%, which was attributed to high impact velocity and severe plastic deformation of the particles during the cold spray process. Hardness of the coating was significantly higher than IN718 substrate. The hardness value in regions closer to the substrate-coating interface is higher than those closer to coating upper surface due to peening effect. While this study focused on hardness of the cold sprayed coatings, it is hoped that further tribological properties will be investigated based on the promising results reported in this paper, since higher hardness is an indicator of better wear resistance.

Acknowledgements. The research was funded by National Research Foundation of Singapore, Rolls-Royce and Nanyang Technological University through research grant ARMS 1.1 (Advanced Metallization Coatings using Cold Spray).

References

1. ASTM International: ASTM E2109−01: Standard Test Methods for Determining Area Percentage Porosity in Thermal Sprayed Coatings, pp. 1–8. ASTM International, West Conshohocken (2014)
2. Chaudhuri, A., Raghupathy, Y., Srinivasan, D., Suwas, S., Srivastava, C.: Microstructural evolution of cold-sprayed Inconel 625 superalloy coatings on low alloy steel substrate. Acta Mater. **129**, 11–25 (2017)
3. Bhushan, B.: Introduction to Tribology, 2nd edn. John Wiley & Sons, West Sussex (2013)

Evaluation of Powders for Cold Spray Coatings Through Study of Single Particle Impact

Rong Gen Neo[1], Nataniel Yong Syn Tham[1,4], Kaiqiang Wu[1,2], Sreerag Puthan Veetil[1], Sung Chyn Tan[3], and Wei Zhou[1,2](✉)

[1] Rolls-Royce@NTU Corporate Laboratory, Nanyang Technological University, 65 Nanyang Drive, Singapore 637460, Singapore
{ronggen.neo,mwzhou}@ntu.edu.sg, wzhou@cantab.net
[2] School of Mechanical and Aerospace Engineering, Nanyang Technological University, 50 Nanyang Avenue, Singapore 639798, Singapore
[3] Rolls-Royce Singapore Pte. Ltd, 1 Seletar Aerospace Crescent, Singapore 797575, Singapore
[4] Advanced Remanufacturing and Technology Center (ARTC), 3 Cleantech Loop, Singapore 637143, Singapore

Abstract. Cold spray is a solid-state powder deposition technique suitable for repair of parts. Use of nominally the same powders from different manufacturers is known to lead to very different porosity levels in the cold spray coatings, so it is important to find an effective method for the powder evaluation. A comparative study was carried out of Ti6Al4V powders from two different manufacturers (plasma atomized powders from AP&C and gas atomized powders from Carpenter Additive). Both powders were deposited to a flat Ti6Al4V substrate using the same cold spray processing parameters, but porosity level was found to be significantly lower in the coatings of the AP&C powders. Both powders were deposited in single layer to make it possible to study single particle impact, and the AP&C powders were found to have 21.1% higher flattening ratio by forming a more pronounced jetting at the particle circumference after impact. The result serves to demonstrate that flattening ratio and jetting after single particle impact can be used as useful indicators of the powder quality.

Keywords: Cold spray · Powder evaluation · Gas atomized powder · Plasma atomized powder · Single particle impact

1 Introduction

Cold spray technology has attracted considerable attention recently due to its capability to repair metallic parts by depositing powders in solid state. This process has been increasingly used in industries. In the aerospace industry, titanium alloys are commonly used. For repairing parts of titanium alloys using cold spray process, Ti6Al4V powders from AP&C are commonly used, perhaps because of its leading position in suppling high quality fine powders. The current study aimed to evaluate Ti6Al4V powders from this company in comparison with nominally the same powders from a different manufacturer (Carpenter Additive, USA). It is found that study of single particle impact may provide useful information for evaluation of powders for cold spray coatings.

Y. Wei and S. Chng (Eds.): INCASE 2021, LNME, pp. 163–167, 2022.
https://doi.org/10.1007/978-981-16-5763-4_36

2 Methods

2.1 Cold Spray Process

Ti6Al4V substrate (Grade 5) with dimensions of 50 × 50 × 4 mm^3 was used. It was ground with 1200 SiC sandpaper and degreased prior to the cold spray process. Two types of Ti6Al4V powders (Grade 23) were used as the feedstock: plasma atomized powder from AP&C and gas atomized powder from Carpenter Additive. The powder diameter was 15–45 μm (D_{10} 16.7 μm, D_{50} 35.5 μm, D_{90} 55.6 μm) for AP&C and 15–53 μm (D_{10} 21.7 μm, D_{50} 40.5 μm, D_{90} 63.3 μm) for Carpenter Additive. The powders were deposited with Impact Spray System 5/11 (Impact Innovations, Germany) with a velocity around 800 m/s. The velocity and diameter of the powders were measured using Tecnar Cold Spray Meter. The substrate was placed on the sample stage that moved horizontally at 500 mm/s during the deposition process. The experimental setup is shown in Fig. 1. Coatings of 2 mm in thickness were deposited for measurement of porosity levels, and single layer deposition was also made for the study of single particle impact.

Fig. 1. Cold spray setup showing the placement of the gun and Ti6Al4V substrate. The substrate is positioned perpendicularly to the gun spray direction.

2.2 Material Characterization

The cold sprayed samples were cut to the required sizes, mounted, ground, and polished to mirror-like finish prior to observation. The deformation of individual particles deposited on the sample and cross-sections of the samples were observed under the laser confocal microscope (Olympus, LEXT OLS4100). The porosity level was measured using the ImageJ software (NIH, USA) and 4 optical images of the coatings were used for measurement at each distance from the coating-substrate interface.

3 Results and Discussion

After the single layer cold spray deposition, study of the single particles under the laser confocal microscope revealed that AP&C particles had pronounced jetting around their

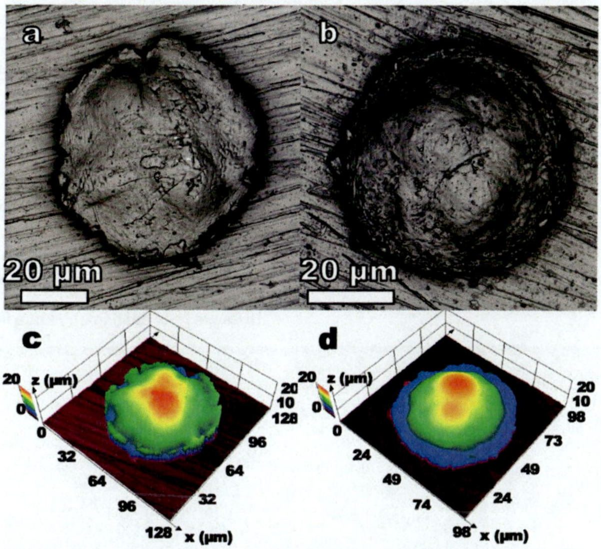

Fig. 2. Laser confocal microscope images showing single particles after the single layer cold spray deposition. (a) and (c) AP&C particle; and (b) and (d) Carpenter Additive particle.

circumferences, and in contrast the Carpenter Additive particles showed no discernible jetting, as shown in Fig. 2.

Further characterization of the particle deformation was done by calculating the flattening ratio, which is defined as the ratio of the diameter to the height of the deformed particle. Five randomly chosen well-deformed particles were scanned to get their profiles and the flattening ratios were determined. AP&C particles were heavily deformed with a 21.1% higher flattening ratio than Carpenter Additive particles, as shown in Fig. 3(a). This is due to different amounts of oxygen in the feedstock powder contributed by different manufacturing processes [1]. Data from the manufacturer mentioned that the plasma atomized powder (AP&C) has a lower oxygen content (0.11 wt.%) than the gas atomized powder (Carpenter Additive) (0.13 wt.%).

In the 2 mm-thick coatings produced, the porosity was found to be lower than 3% for both types of powders. AP&C has a lower porosity than Carpenter Additive powder at a distance 0.5, 1.0, and 1.5 mm away from the coating-substrate interface, as shown in Fig. 3(b). At 0.5 mm distance, results show a 45.2% lower porosity measured for AP&C powder. Furthermore, from the optical images in Figs. 3(c) and (d), the overall porosity of AP&C coating is visually lower than Carpenter Additive coating.

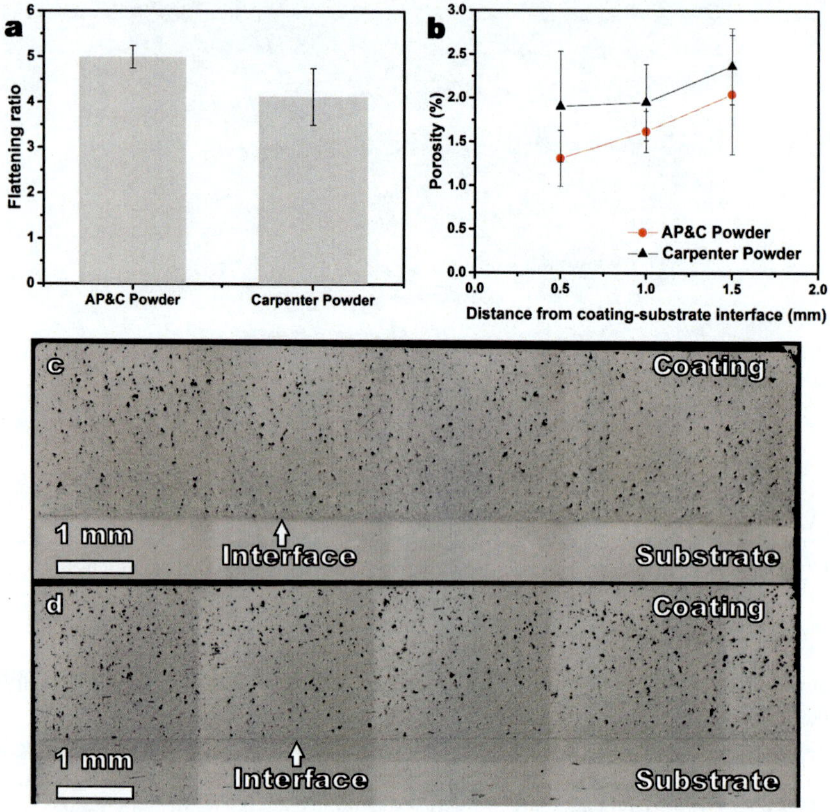

Fig. 3. (a) Flattening ratio of the two types of powders. (b) Porosity of the 2 mm-thick cold spray coating showing AP&C has a lower porosity. Cross-sections of (c) AP&C and (d) Carpenter Additive coatings.

4 Conclusions

Powder of same material produced by different manufacturers were deposited on the same substrate for analytical study. This study found that even powders made of same material has a different deformation behavior. Powders manufactured via the plasma atomization process deformed 21.1% more via the creation of jet around the circumference of the particle after impact and this results in a lower coating porosity. The study provides a possible method to select powders for cold spray applications.

Acknowledgements. The research was funded by National Research Foundation of Singapore, Rolls-Royce and Nanyang Technological University through research grant ARMS 1.1 (Advanced Metallization Coatings using Cold Spray).

Reference

1. Bhattiprolu, V.S., Johnson, K.W., Ozdemir, O.C., Crawford, G.A.: Influence of feedstock powder and cold spray processing parameters on microstructure and mechanical properties of Ti-6Al-4V cold spray depositions. Surf. Coat. Technol. **335**, 1–12 (2018)

Post-processing of Cold Spray Coatings via Rapid Induction Heating

Yongjing Yang[1], Kaiqiang Wu[1], Aprilia Aprilia[2], Sung Chyn Tan[3], and Wei Zhou[1,2](✉)

[1] School of Mechanical and Aerospace Engineering, Nanyang Technological University, 50 Nanyang Avenue, Singapore 639798, Singapore
wzhou@cantab.net, mwzhou@ntu.edu.sg
[2] Rolls-Royce@NTU Corporate Laboratory, Nanyang Technological University, 65 Nanyang Drive, Singapore 637460, Singapore
aprilia@ntu.edu.sg
[3] Rolls-Royce Singapore Pte. Ltd, 1 Seletar Aerospace Crescent, Singapore 797575, Singapore

Abstract. Cold spray is an additive manufacturing process increasingly used for repair and restoration of damaged components. A common problem with the technology is the high level of porosity in the as-sprayed coatings. Effort has been made in the past to reduce the porosity level by post-processing methods such as heat treatment in vacuum furnace. However, the conventional heat treatment is a slow process which typically takes a few hours. The prolonged heating and cooling cycles reduce productivity and could cause some adverse effects on the components. The current research aimed to develop a rapid technique for post-processing of cold spray coatings. Cold spray was carried out to coat multiple layers of CoNiCrAlY powders onto Inconel 718 substrate. The coatings were then post-processed by induction heating at 800, 900, 1000 and 1100 °C. The experimental results show that induction heating reduced the porosity level significantly in a short period of merely 10 min. The effectiveness of induction heating is attributed to the preferential electrical resistance heating of regions with pores. The study demonstrates the potential application of induction heating for rapid post-processing of cold spray coatings.

Keywords: Cold spray · Post-processing · Induction heating · Porosity

1 Introduction

Cold spray is a solid-state additive manufacturing process to deposit coatings on substrates with minimal heat input and thermally induced distortion. During the cold spray process, the feedstock particles hit the substrate at a supersonic speed to cause bonding through severe plastic deformation. However, the inter-particle bonding is usually limited in the cold spray coatings because the atomic diffusion is relatively restricted, resulting in high level of porosity.

Considerable effort has been made in the past decade to reduce the porosity level by post-processing, but the post-processing is mainly conducted by conventional heat

Y. Wei and S. Chng (Eds.): INCASE 2021, LNME, pp. 168–172, 2022.
https://doi.org/10.1007/978-981-16-5763-4_37

treatment in vacuum furnace, which is a slow process typically taking a few hours [1]. The current research aimed to develop a rapid induction heating technique for post-processing of cold spray coatings.

2 Methods

2.1 Cold Spray Process

The substrate was an Inconel 718 plate of 8 mm in thickness. It was ground using SiC paper and ultrasonically cleaned prior to the cold spray process. The feedstock was gas atomized CoNiCrAlY powder of ~45 μm in mean size supplied by Oerlikon Metco.

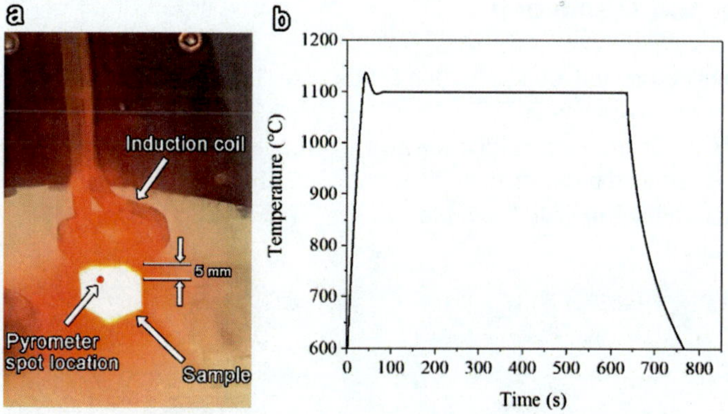

Fig. 1. (a) Photograph showing experimental setup for the induction heating process; (b) A typical temperature profile achieved (1100 °C, holding time 10 min).

High pressure cold spray deposition was carried out using an Impact Spray System 5/11 (Impact Innovations GmbH, Germany) through a SiC nozzle. The CoNiCrAlY powder was deposited onto the substrate at a feed rate of about 45 g/min. Pure nitrogen as propellant gas was preheated to 1000 °C at about 4.5 MPa gas pressure. The traverse speed of the nozzle was 500 mm/s. The coating consisted of 10 layers in total.

2.2 Induction Heating

The induction heating equipment used (GH Induction Atmospheres, USA) consisted of a high-frequency generator with a maximum power of 12 kW and a frequency of 107 kHz. A vacuum bell jar system was used to prevent surface oxidation. The vacuum pressure inside the bell jar was set to be at around 5×10^{-4} torr. For each experimental run, the sample with the coating on its top surface was positioned directly under the induction coil of 30 mm in diameter, the stand-off distance was set at 5 mm, and an infrared pyrometer was used for the closed-loop temperature control with the sensing spot placed on the coating surface, as shown in Fig. 1(a). The samples were heated to 800, 900, 1000 and 1100 °C respectively for a fixed period of 10 min. After heating, the samples were cooled down in vacuum for 5 min and then in Ar gas for another 5 min before being cooled to room temperature in the open atmosphere. A typical induction heat treatment profile is shown in Fig. 1(b).

2.3 Materials Characterization

The samples were mounted, ground and polished to mirror-like finish prior to the observation. The cross-sections of the samples were observed under the optical microscope (Olympus MX40) and field emission scanning electron microscope (JEOL-7600). The porosity level was measured using the ImageJ software (NIH, USA) and at least 6 optical images were used for measurement of each sample.

3 Results and Discussion

Cross-sections before and after induction heating are shown in Fig. 2. Severe pores were found in the as-sprayed coating (Fig. 2(a)). Figures 2(b–e) show that the porosity level was reduced significantly after induction heating from 800 °C to 1100 °C. Quantitatively, the porosity is reduced by more than 90% from 2.2% in the as-sprayed sample to below 0.2% in all the induction heated samples, as shown in Fig. 2(f).

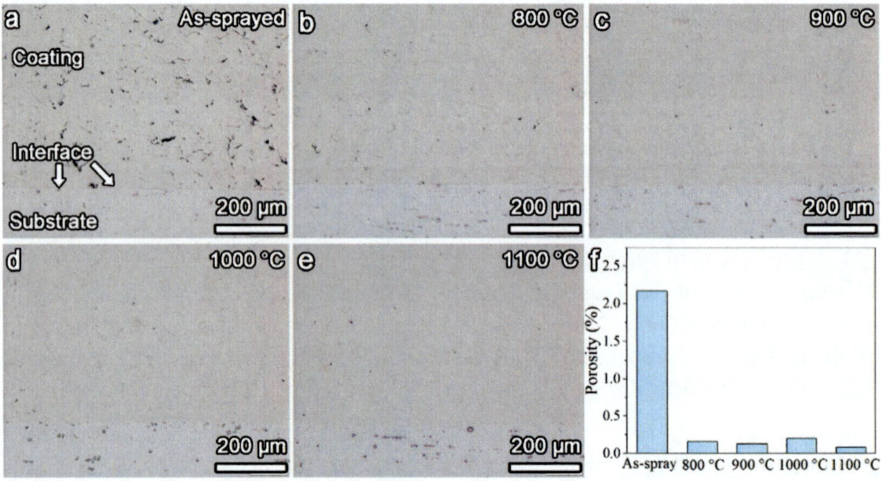

Fig. 2. Optical micrographs showing cross-sections of the cold sprayed samples: (a) As-sprayed; (b)–(e) Induction heated at four different temperatures; and (f) Quantitative results of the porosity.

To study the pores further, back-scattered electron microscopy was conducted. Figures 3(a, d) show large pores due to the poor inter-particle bonding in the as-sprayed coating. After induction heating at 800 °C, some pores remained and they were still observed to be inter-particle pores (Fig. 3(b, e)). When the inducting heating temperature was increased to 1100 °C, almost pore-free microstructure formed, as shown in Figs. 3(c, f).

During induction heating, the eddy current cannot go through the pores, which have almost infinite electrical resistivity. This would lead to current detouring and crowding around the pores, which would result in preferential Joule heating to cause closure of the pores. The mechanisms require further study, but a previous study by Song et al. shows crack healing effect triggered by electropulsing stimuli [2], leading support to the proposed mechanisms.

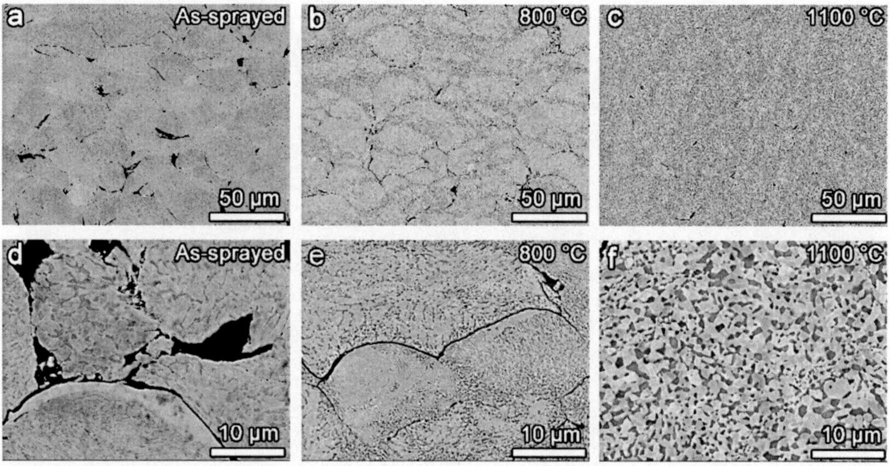

Fig. 3. Back-scattered electron images showing the particle bonding of cold sprayed samples: (a, d) As-sprayed; (b, e) Induction heating at 800 °C; and (c, f) Induction heating at 1100 °C.

4 Conclusions

The cold spray coatings of CoNiCrAlY on Inconel 718 substrate were post-processed by induction heating at different temperatures ranging from 800 °C to 1100 °C. It is revealed that the microstructural improvement was significant through over 90% porosity reduction and enhancement of inter-particle bonding. Induction heating is demonstrated to be an effective tool for rapid post-processing of cold spray coatings. It should be noted that the research focused on porosity reduction only, and it is hoped that the promising results obtained will stimulate interest to study other important engineering issues such as enhancement of interfacial bonding between cold spray coating and substrate.

Acknowledgements. The research was funded by National Research Foundation of Singapore, Rolls-Royce and Nanyang Technological University through research grant ARMS 1.2 (Repair and Restoration of Airfoils by Induction Brazing).

References

1. Zhou, H.X., et al.: Pores structure change induced by heat treatment in cold-sprayed Ti6Al4V coating. J. Therm. Spray Technol. **28**, 1199–1211 (2019)
2. Song, H., Wang, Z.J., He, X.D., Duan, J.: Self-healing of damage inside metals triggered by electropulsing stimuli. Sci. Rep. **7**, 1–11 (2017)

Corrosion Resistance of Unsealed Plasma Sprayed Coating Under Two Different Accelerated Exposures

Yong Teck Tan[1](✉) and Poh Koon Aw[2]

[1] Electrochemical Process Section, Surface Technology Group, Singapore Institute of Manufacturing Technology, A*STAR Research Entities, 2 Fusionopolis Way, #08-04, Innovis, Singapore 138634, Singapore
tanyongt@SIMTech.a-star.edu.sg

[2] Physical Deposition Process Section, Surface Technology Group, Singapore Institute of Manufacturing Technology, A*STAR Research Entities, 2 Fusionopolis Way, #08-04, Innovis, Singapore 138634, Singapore
pkaw@SIMTech.a-star.edu.sg

Abstract. Plasma sprayed coatings are widely used in many industries. These coatings are porous in nature and require sealing for good barrier properties. To investigate the intrinsic corrosion resistance of unsealed plasma sprayed coating, this work explored the corrosion behaviour of an unsealed plasma sprayed coating system consisting of Al_2O_3-TiO_2 (AT) coating with a NiCr bond coat on AISI 1045 carbon steel, under two different types of accelerated exposures: continuous salt fog (ASTM B117) and cyclic exposure (ISO 14993). A comprehensive approach was used for investigation on different control samples and mechanically damaged samples, including visual examination, SEM and EDX mapping, and electrochemical experiments. It was found that the NiCr coating by itself does not provide significant corrosion protection. The two-coat system protects the carbon steel effectively against corrosion by a barrier mechanism. Electrochemical results show that corrosion resistance of undamaged two-coat (AT + NiCr) system increases ten-fold after exposure. This improvement in corrosion resistance is hypothesized to be due to a self-sealing effect in which iron corrosion products reduce the coating porosity.

Keywords: Carbon steel · Metal coating · Ceramic coating · Accelerated exposure · Electrochemical test

1 Introduction

Plasma sprayed Al_2O_3-based and Al_2O_3-TiO_2-based coatings have good wear and corrosion resistances and are used for a wide range of industrial applications [1–6]. The microstructure of thermal sprayed coatings contains inherent porosity, which is detrimental to the mechanical strength, wear resistance and corrosion resistance of the sprayed coating [7]. To improve corrosion resistance, plasma sprayed coatings usually require sealing for better barrier properties.

Y. Wei and S. Chng (Eds.): INCASE 2021, LNME, pp. 173–176, 2022.
https://doi.org/10.1007/978-981-16-5763-4_38

To investigate the intrinsic corrosion resistance of unsealed plasma sprayed coating, in this work we explore the corrosion behaviour of an unsealed plasma sprayed coating system consisting of Al_2O_3-TiO_2 (AT) coating with a NiCr bond coat on AISI 1045 carbon steel, under two different types of accelerated exposure, and also under the condition of a mechanical damage reaching to the steel substrate.

2 Methods

AISI 1045 medium carbon steel was used as the substrate metal. Fused and crushed Al_2O_3 and TiO_2 powder of a proprietary composition, particle size range 20–50 μm, was used for the Al_2O_3-TiO_2 (AT) coating layer (300–350 μm thick). Spherical 80 wt.% Ni-20 wt. % Cr powder, particle size range 15–50 μm, was used for the NiCr coating layer (70–100 μm thick). The NiCr layer is the bond coat between the steel and AT layer for two-coat system. The plasma spray machine used was an Oerlikon Metco F4 plasma spray system using Ar-H_2 plasma gas. The plasma spray process parameters used here are similar to typical plasma spray process conditions found in literature [8], and were fine-tuned in a proprietary study to minimize the coating porosity. The coated region was 40 mm × 25 mm for all samples. Two types of coating samples were prepared: undamaged coating samples and artificially damaged coating samples, where a linear cut of 0.65 mm width and 13 mm length exposed the carbon steel substrate.

Two different types of accelerated corrosion exposure were used. Total exposure period was 1008 h. The first is a continuous salt fog of 5 wt.% NaCl at 35 °C, according to ASTM B117–16 [9]. The second is a cyclic exposure according to ISO 14993 [10], involving repetitions of 8 h-long cycles, with each cycle consisting of three stages: 2 h of salt mist (5 wt. % NaCl, 35 °C), followed by 4 h of drying and 2 h of wet conditions. Photos for visual assessment were taken with a Nikon D5500 camera in a colour assessment cabinet under controlled lighting conditions.

Scanning electron microscope (SEM) imaging and Energy Dispersive X-ray spectroscopy (EDX) mapping were conducted using a JEOL JSM-IT300 SEM equipped with an Oxford Instruments X-MaxN EDX detector. The electrochemical tests in 0.6 M NaCl solution were conducted using a three-electrode set-up (SCE as reference electrode and graphite rod as counter electrode) and a Gamry Reference 600 potentiostat.

3 Results and Discussion

Figure 1 shows the visual images of undamaged coating samples at the start and end of the 1008 h exposures. Undamaged two-coat system (AT + NiCr) offered good corrosion protection under both accelerated exposure conditions for 1008 h, excluding edge defects. Some speckling occurred for accelerated cyclic exposure samples, which could be due to localized rusting or precipitation of surface runoff from edge defects. For artificially damaged two-coat samples under continuous salt fog exposure, there was significant red rust formation on the surface by Day 7 (168 h). For artificially damaged two-coat samples under cyclic exposure, the formation of rust within the damage zone reduced the extent of red rust formation on the external surface, even on Day 42 (1008 h). Undamaged NiCr coating by itself did not protect the carbon steel substrate effectively against corrosion. Significant red rust (covering >50% of the exposed area) was observed by Day 7 under ASTM B117 continuous salt fog exposure and by Day 14 under ISO 14993 cyclic exposure. The corrosion behaviour of undamaged and damaged NiCr coating samples were similar, i.e. even in the absence of any artificial defect, the inherent porosity of the NiCr bond coat results in significant red rust formation, like that of bare carbon steel. This indicates that the NiCr bond coat by itself is not sufficient for corrosion protection.

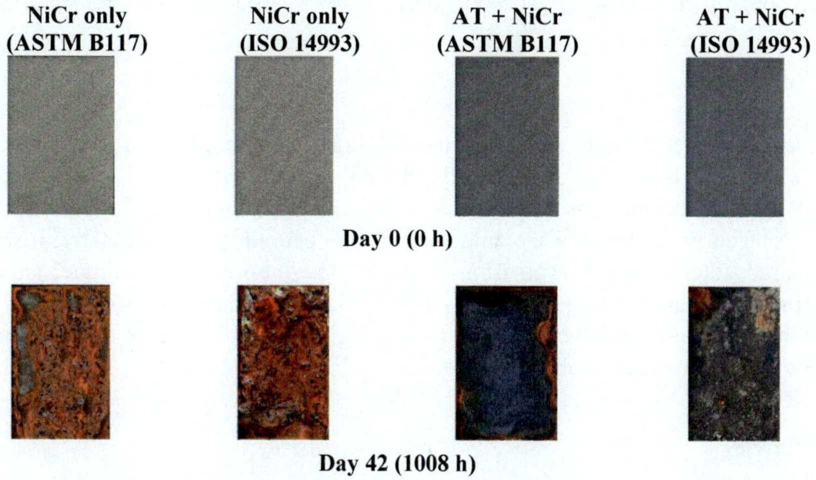

Fig. 1. Visual images of coating samples under accelerated exposure conditions according to ASTM B117 and ISO 14993 at the start and end of 1008 h exposure.

The samples were further analyzed with cross-sectional SEM and EDX mapping. Under both accelerated exposure conditions, the coating layers of undamaged AT + NiCr coating remained intact with no deformation or cracking and no visible corrosion product formation at all interfaces.

For artificially damaged AT + NiCr coating sample under accelerated cyclic exposure conditions, rust formation within the damage zone helped to plug up the damage zone, in contrast to exposure in continuous salt fog, where the damage zone served as a path of least resistance for the transport of ferrous cations from corrosion sites at the exposed carbon steel region to the external environment, resulting in large amount of red rust seen on the external coating surface.

Thus, under continuous salt fog exposure conditions, precipitation of iron corrosion products occur more slowly (sufficient time for ferrous cations to travel to external coating surface) while under cyclic exposure conditions, the precipitation of iron corrosion products occur more quickly (precipitation within the coating).

Electrochemical impedance spectroscopy (EIS) experiments were conducted to determine the total resistance to corrosion. The average total resistance (R_{tot}) for fresh NiCr and fresh AT + NiCr coating samples were 0.92 kΩ cm^2 and 1.5 kΩ cm^2 respectively. For the ASTM B117-exposed AT + NiCr samples, the average R_{tot} was around

15 kΩ cm^2 which is one order of magnitude higher than the fresh two-coat samples, indicating that the corrosion resistance increased during the exposure. For the ISO 14993-exposed AT + NiCr samples, though the R_{tot} cannot be calculated directly due to the presence of a diffusion element, the impedance magnitude was higher than that of the fresh AT + NiCr coating at all tested frequencies which imply that the R_{tot} will be higher than the fresh coating.

The one order of magnitude increase in corrosion resistance during exposure is hypothesized to be due to a self-sealing effect in which Fe corrosion products reduce the coating porosity of the coating, especially near the NiCr-carbon steel substrate interface.

4 Conclusions

The plasma sprayed NiCr bond coat by itself did not increase the corrosion resistance of the carbon steel substrate significantly. The AT + NiCr two-coat system had good corrosion resistance via a barrier protection effect. Electrochemical test results showed that the total corrosion resistance of undamaged two-coat samples increased by an order of magnitude after accelerated corrosion exposures. The one order of magnitude increase in corrosion resistance during exposure is hypothesized to be due to a self-sealing effect in which Fe corrosion products reduce the coating porosity of the coating, especially near the NiCr-carbon steel substrate interface.

References

1. Xie, Y., Hawthorne, H.M.: Wear mechanism of plasma-sprayed alumina coating in sliding contacts with harder asperities. Wear **225–229**, 90–103 (1999)
2. Lin, X., Zeng, Y., Ding, C., Zhang, P.: Effects of temperature on tribological properties of nanostructured and conventional Al2O3–3 wt.% TiO2 coatings. Wear **256**(11–12), 1018–1025 (2004)
3. Wang, Y., Tian, W., Zhang, T., Yang, Y.: Microstructure, spallation and corrosion of plasma sprayed Al_2O_3-13%TiO_2 coatings. Corros. Sci. **51**, 2924–2931 (2009)
4. Tian, W., Wang, Y., Zhang, T., Yang, Y.: Sliding wear and electrochemical corrosion behavior of plasma sprayed nanocomposite Al_2O_3-13%TiO_2 coatings. Mater. Chem. Phys. **118**, 37–45 (2009)
5. Habib, K.A., Saura, J.J., Ferrer, C., Damra, M.S., Giménez, E., Cabedo, L.: Comparison of flame sprayed Al2O3/TiO2 coatings: their microstructure, mechanical properties and tribology behavior. Surf. Coat. Technol. **201**, 1436–1443 (2006)
6. Heimann, R.B: Plasma Spray Coating; 2nd edn. Wiley-VCH, Weinheim (2008)
7. Leivo, E.M., Vippola, M.S., Sorsa, P.P.A., Vuoristo, P.M.J., Mäntylä, T.A.: Wear and corrosion properties of plasma sprayed Al_2O_3 and Cr_2O_3 coatings sealed by aluminum phosphates. J. Therm. Spray Technol. **6**, 205–210 (1997)
8. Xiao, Y., Shao, F., Zhuang, Y., Ni, J., Zhao, H., Tao, S.: Comparative study of the corrosion resistance of air-plasma-sprayed Ca_2SiO_4 and Al_2O_3 coatings in salt water. Coatings **8**, 1–15 (2018)
9. ASTM B117–16: Standard Practice for Operating Salt Spray (Fog) Apparatus, pp. 1–12 (2016)
10. ISO 14993:2001 Corrosion of metals and alloys – Accelerated testing involving cyclic exposure to salt mist, “dry” and “wet” conditions, pp. 1–10 (2001)

High Performance Cool Coatings

Siming Dong, Alexander van Herk, and Satyasankar Jana(✉)

Institute of Chemical and Engineering Sciences, A*STAR, 1 Pesek Road, Singapore, Singapore
satyasankar_jana@ices.a-star.edu.sg

Abstract. Cool coatings are coatings with high solar reflectivity. We hereby report the synthesis and application of polymer-encapsulated pigment to improve cooling performance of cool coatings and synthesis and application of novel dirt-repellent fluorinated core-shell particles (FCSP) for the improvement of TSR-durability. Waterborne cool coating formulations produced using PMMA encapsulated TiO_2 showed ~8% improvement in total solar reflectance (TSR) and 10% improvement in NIR reflectivity. Cool coating formulations using FCSP showed up to 30% more dirt-repellent properties in compare to unmodified commercial cool coating.

Keywords: Cool coatings · Solar reflectivity · Dirt-repellency · Waterborne coating

1 Introduction

Coatings with high total solar reflectance (TSR) are often termed as "cool coatings". Cool coatings, when applied to the exterior surfaces of the buildings, reduce surface heat build-up as it reflects most of the incident solar radiation and, therefore, reduce building heat gain during the daytime. This ultimately reduces air-con energy consumption required for the building cooling. Application of cool coating over the large surface areas can cause reduction in localized ambient temperature during solar peak hours and mitigate urban heat island effect (UHI) during evening and night [1, 2]. Cool coatings are also mentioned to have potential to mitigate global warming [3] due to the fact that reflected solar light can go back to space by passing through light-transparent atmosphere. Therefore, this passive cooling technology is indirectly potential to reduce CO_2 emission too. Rutile TiO_2 is most widely used white pigment to formulate cool coatings due to its high solar reflectivity additional to its contribution towards excellent hiding power and whiteness. Uniform distribution of TiO_2 in the coating and TiO_2-TiO_2 pigment spacing in the coating film play important roles on the performance of the cool coatings. Additional to the initial cooling performance or TSR of the cool coatings, the durability of cooling performance over the long term is also a matter of concern for the real application of cool coatings. Long-term durability of cooling performance of cool coatings often drastically reduced due to environmental dirt-accumulation.

This oral presentation will highlight general science and technology of cool coatings, improvement of cool coating performance by the use of polymer-encapsulated TiO_2 to avoid TiO_2 crowding and by controlling TiO_2-TiO_2 pigment spacing, and the formulation of durable cool coatings with the use of dirt-repellent fluorinated core-shell (FCSP) particles.

Y. Wei and S. Chng (Eds.): INCASE 2021, LNME, pp. 177–180, 2022.
https://doi.org/10.1007/978-981-16-5763-4_39

2 Results and Discussion

Polymer encapsulated rutile TiO_2 was synthesized via an emulsion polymerization method and purified as described in the literature [4]. Cool coatings using commercial TiO_2 (R706) and polymer encapsulated TiO_2 were formulated (S01-S10) using specific amount of latex binder (Primal, AC261) and small amount of coatings additives [4]. Finally, the coatings films were casted on the stainless steel substrates (cast area 6 cm × 6 cm) and dried fully before solar reflectivity measurement (Perkin-Elmer Lambda 900) and other characterizations. The film thickness of coatings was controlled between 90–110 μm (measured by an Elcometer thickness gauge).

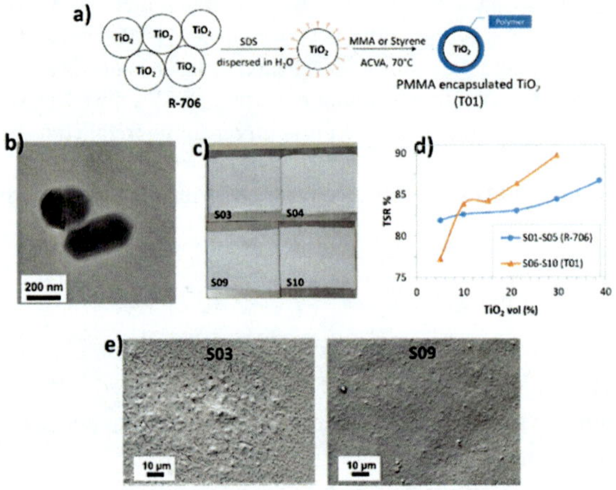

Fig. 1. a) Synthesis scheme of polymer encapsulation of TiO_2, b) SEM image of encapsulated TiO_2 (T01) showing PMMA shell, c) optical images of coatings formulated using commercial un-encapsulated TiO_2, R-706 (S03-S04) and PMMA-encapsulated TiO_2, T01 (S09-S10) and other additives, d) variation of TSR with the different loading of R-706 and T01, and e) SEM images of a coatings produced using R-706 (S03) and T01 (S09).

PMMA encapsulated rutile TiO_2 (T01) with the shell thickness of 30–40 nm was successfully synthesized as seen in Fig. 1a–b. Cool coatings formulated using T01 pigment showed substantial increase of total solar reflectance (TSR) above the dosage of 10 vol% TiO_2 (Fig. 1d). Cool coating S10 formulated with 29.7 vol% TiO_2 using T01 showed highest TSR value and approximately 8% higher than the coating formulated using un-encapsulated commercial TiO_2 (S04). This increase in TSR (and also NIR reflectivity) of coatings formulated using encapsulated TiO_2 is due to the improved dispersion of encapsulated particles in coatings, increase in TiO_2-TiO_2 spacing that improves scattering of light. Other influencing factor may include the multiple refractive index (RI) mismatch (matrix-shell-particle), which also known to improve light scattering. The surface of the coating film of S03 produced using commercial un-encapsulated TiO_2 (R-706) showed marginally rougher at high magnification SEM image (Fig. 1e) than the surface of the coating film (S09) produced using PMMA-encapsulated TiO_2, T01. Similar results were also obtained from the surface roughness analyses by AFM. This rough surface may indicates aggregation of TiO_2 in the film, which can also affect other coatings properties like chalk resistance and gloss retention in the long term. All coatings formulated and used for this study possessed approximately 85% emissivity. Cool coatings formulated using PMMA encapsulated TiO_2 possessed better cooling properties in lab under IR light irradiation and outside under natural irradiation.

Fluorinated core shell particles (FCSP) with poly(methyl methacrylate) (PMMA) core and fluoro methacrylate polymer shell were synthesized by a patented emulsion polymerization method [5]. FCSP particles were directly mixed with a commercial cool coating, casted on the stainless steel substrates (cast area 6 cm × 6 cm) and then dried fully before the measurement of dirt-repellent properties via a lab based anti-dirt measurement.

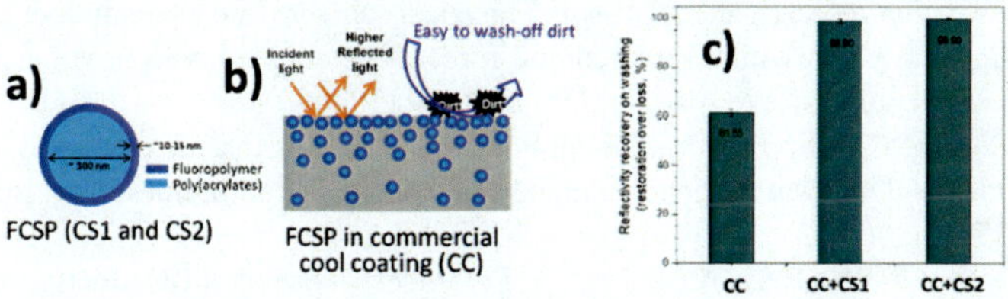

Fig. 2. a) Structure of fluorinated core-shell particle (FCSP), b) functionality of FCSPs when present in waterborne coatings and c) anti-dirt properties of a commercial cool coating (CC) and FCSP mixed commercial cool coatings (CC+CS1 and CC+CS2). CS1 and CS1 are FCSPs with different fluoropolymer shell.

FCSPs of 300 nm size were successfully synthesized via an emulsion polymerization method. FCSPs were proven to be well compatible with the commercial waterborne cool coatings supported by different characterization methods. Addition of FCSP to the coating helped to improve both hydrophobicity and oleophobicity as observed by the increase in water and oil contact angles. For anti-dirt test, at first a specific amount of dry natural air-con dirt was sprayed over the coated coupons and rested for couple of days before the coupons were subjected to water spray wash simulating rain in Singapore. The amount of dirt before and after water spray was quantified by the measurement of light reflectivity as an indirect method to quantify dirt. FCSP mixed commercial coatings were proven to improve ~30% more dirt-pic up resistance property (Fig. 2c) in compare to pure commercial coating. TSR-durability of FCSP modified commercial cool coatings were validated by the outdoor test bedding for one year without much degradation of TSR and other coatings properties.

3 Conclusion

Total solar reflectance (TSR) and the cooling performance of cool coatings can be significantly improved by the use of polymer-encapsulated TiO_2 to avoid TiO_2 crowding and by controlling TiO_2-TiO_2 pigment spacing within the coating. Use of fluorinated core-shell (FCSP) particles to waterborne cool coatings improves dirt-repellent properties of cool coatings, which will enhance TSR-durability of cool coatings.

Acknowledgement. The authors would like to acknowledge their collaborators Prof. Wan Man Pun from Nanyang Technological University, Singapore and Mr Christopher Arthur Crawley from SkyCool, Singapore for their support on different characterizations and for providing commercial cool coating.

References

1. Santamouris, M.: Cooling the cities–a review of reflective and green roof mitigation technologies to fight heat island and improve comfort in urban environments. Sol. Energy **103**, 682–703 (2014)
2. Mandi, Z., Yongping, L., Zhang, X., Donthu, E.V.S.K.K., Bing Feng N., Man Pun W.: Sensitivity study of weather research and forecasting physical schemes and evaluation of cool coating effects in singapore by weather research and forecasting coupled with urban canopy model simulations. J. Geophys. Res.: Atmos. **125**(13), 1–19 (2020)
3. Rossi, F., Cotana, F., Filipponi, M., Nicolini, A., Menon, S., Rosenfeld, A.: Cool roofs as a strategy to tackle global warming: economical and technical opportunities. Adv. Build. Energy Res. **7**(2), 254–268 (2013)
4. Dong, S., Quek, J.Y., Van Herk, A.M., Jana, S.: Polymer encapsulated TiO_2 for the improvement of NIR reflectance and total solar reflectance of cool coatings. I&EC Res. **59**, 17901 (2020)
5. Jana, S., Wan M.P., Arthur Crawley, C., Barik, S., Zihan, P., Van Herk, A.: US20190375946A1, Polymer dispersion (2019)

Influences of Porosity on the *In-Vitro* Cell Responses of Cold Sprayed Ti-Ta Composites

Guang Zeng[1(✉)], Saden H. Zahiri[2], Stefan Gulizia[2], Xiao-Bo Chen[1], and Ivan Cole[1]

[1] RMIT University, Melbourne, VIC 3000, Australia
guang.zeng@student.rmit.edu.au, ivan.cole@rmit.edu.au

[2] Commonwealth Scientific and Industrial Research Organisation (CSIRO), Clayton, VIC 3168, Australia

Abstract. Surface topography as one of the most influential factors in determining the complex processes of cell responses in tissue engineering has been widely investigated. This study explores contributions of pososity upon cold sprayed Ti-Ta composite to cell responses. *In-vitro* cell culture was firstly discussed the interaction of cells to the polished counterparts with a moderate porosity and with low porosity of Ti-Ta composites, respectively. The significantly enhanced cell growth rate on the polished surface of Ti-Ta composite with moderate porosity compared to the polished surfaces with low porosity was observed during the initial cell culture period (0.5 h; 1 h; 3 h and 6 h). This research indicates cold spray manufacturing could be utilized to produce porous structures with benefits to enhance the osseointegration for biomedical applications.

Keywords: Cold spray · Ti-Ta composite · Porosity · Cell response

1 Introduction

Biomaterial properties, especially its surface characteristics play a crucial role in successful acceptance and integration of bioimplants. Generally, surface topography can provide the guidance to alter cytoskeleton organization, migration and proliferation of cells and therefore allows modifying the desired biomaterials surface to adapt specific orthopedic applications [1]. Porous structures as one of the topographical factors has been widely accepted as effective methods to match the elastic modulus of metal implants with the human bones [2]. However, these surface modification methods present low efficiency due to the inaccuracy in controlling the surface topography such as the porosity proportions [3]. Further to this, there are limited observations related to the mechanism of porosity in controlling cells' activities so this mechanism remains poorly recognized and explained [4].

Cold spray additive manufacturing provides a promising solution in fabrication of porous structures with various benefits such as low process-temperature and the solid-state deposition process [5, 6]. In this study, moderate-porous and low-porous of Ti-Ta composite structures were fabricated through cold spray additive manufacturing. *In-vitro*

Y. Wei and S. Chng (Eds.): INCASE 2021, LNME, pp. 181–184, 2022.
https://doi.org/10.1007/978-981-16-5763-4_40

cell culture was further discussed the interaction of cell response to the polished surfaces with moderate porosity and low porosity, respectively. Results verified the enhanced cell growth rate on cold sprayed moderate-porous structures compared to the low-porous ones and might result from the surface potential variations generated from the porosity. Outcome of this study suggested cold spray additive manufacturing would be utilized in surface topographical modifications with benefits to enhance the osseointegration for biomedical applications.

2 Methods

2.1 Fabrication of Porous Structures

The commercial cold spray system (Impact Innovations, Germany) was equipped to deposite Ti-Ta composite using a mixture of commercially pure Ti powders (Kimet, China) and Ta powders (AMPERIT 151, Starck®) as feedstook with the mass ratio of 7:3 for Ti-Ta composite. A convergent-divergent nozzle (SiC Nozzle Type 24) was applied to accelerate feedstock powders to supersonic speed using nitrogen. Moderate porous Ti-Ta structure was deposited at 400 °C and 4 MPa gas pressure under traverse speed of 20 g/min with 30 mm stand-off distance. Low porous Ti-Ta structure was deposited at 900 °C and 5 MPa gas pressure under traverse speed of 20 g/min with 35 mm stand-off distance.

2.2 *In-Vitro* Biocompatibility

In-vitro biocompatibility testing on all specimens after autoclaving were seeded with an equal density of GPE86 (ATCC®) mouse fibroblast cells (0.1 million/mL × 500 μL) for initial cell culture periods (0.5 h; 1 h; 3 h; 6 h), respectively. Cultures were maintained at 37 ± 0.5 °C under 5% CO_2 and 95% humid atmosphere. After rinsed with phosphate buffered solution (PBS) and fixed using 4% paraformaldehyde for 30 min, the fixed cells on all the specimens were subsequently dehydrated in ethanol series (20, 40, 60 and 80%) for 1 h and stored in 90% ethanol. Cell morphologies were observed under field emission scanning electron microscope (FE-SEM, Zeiss Merlin FEG SEM). Surface potential measurement was carried out by scanning kelvin probe force microscopy (SKPFM, Cypher ES, Asylum Research).

3 Results and Discussion

3.1 Cell Response to the Polished Surfaces with Moderate and Low Porosity

Figure 1 illustrated the cell morphologies on moderate-porous and low-porous of Ti-Ta composite coatings after initial cell incubation period (0.5 h; 1 h; 3 h; 6 h). It can be observed the successful cell growth on both moderate-porous and low-porous surfaces. Figure 1a showed the interaction of cells to the porosity (marked by arrows), which is excluded in Fig. 1b due to the low-porous surface features. The determined cell (attachment) fractions as a function of time are summarized in Fig. 2. As shown, the cell (attachment) fractions on moderate-porous Ti-Ta composite are significantly ($p = 0.0019$; $p < 0.0001$) higher than that on low-porous ones and could achieve 20% higher cell fraction during each initial cell culture period. These results indicate the cell growth rate is higher on moderate-porous Ti-Ta composite compared to the low-porous ones, which might be beneficial to enhancing the osseointegration for biomedical applications.

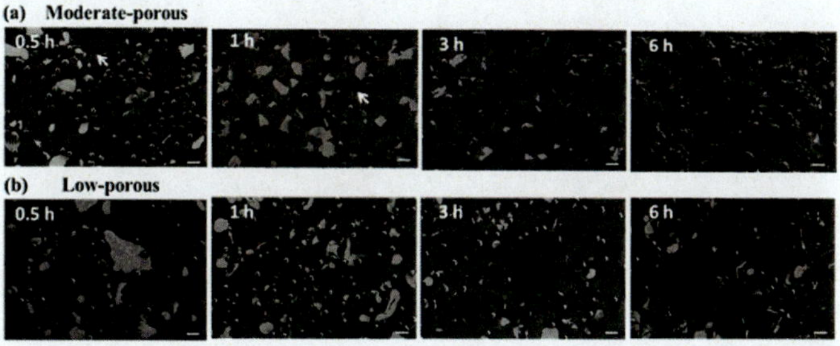

Fig. 1. FE-SEM images of GPE86 cell morphologies on moderate-porous and low-porous cold sprayed Ti-Ta composite coating after 0.5 h; 1 h; 3 h and 6 h incubation period. (a) cell growth on moderate-porous Ti-Ta composite, porosity was indicated by arrow; (b) cell growth on low-porous Ti-Ta composite. Scale bar = 20 μm.

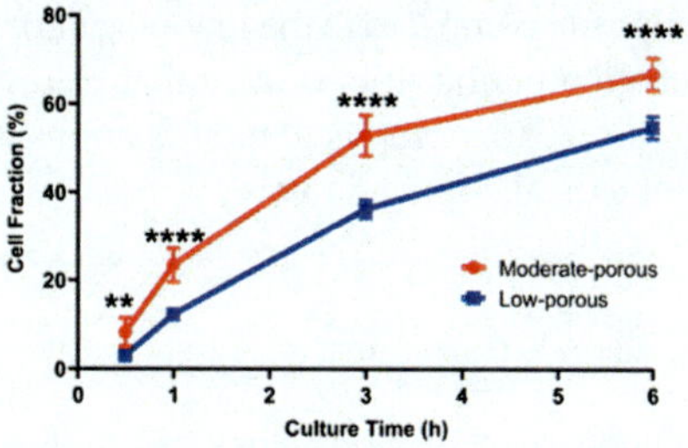

Fig. 2. Cell (attachment) fractions as a function of time for moderate-porous and low-porous Ti-Ta composite coating. Error bars indicate standard deviations and "*" indicates statistically significant differences (determined by GraphPad prism, two-way ANOVA. **$p = 0.0019$; ****$p < 0.0001$).

3.2 Surface Potential Measurement on Cold Sprayed Ti-Ta Composite

Figure 3 shows the SKPFM scanning on the polished moderate-porous Ti-Ta composite with the aim to identify the potential distribution of the porosity that may impact on cell growth. The 3D and 2D surface morphology shown in Fig. 3a and 3b confirmed the apparent depth of the measured porosity. Moreover, compared to the surface potential distributions shown in Fig. 3c and 3d, the significant surface potential variations generated from the porosity were observed. It could also be identified the relatively higher surface potential of Ti matrix compared to Ta particle. These findings revealed the significantly higher measured surface potential existed in the porosity compared to the Ti matrix and Ta particle, suggesting the remarkable surface potential significance generated from moderate-porous Ti-Ta composite that might influence the *in-vitro* cell response.

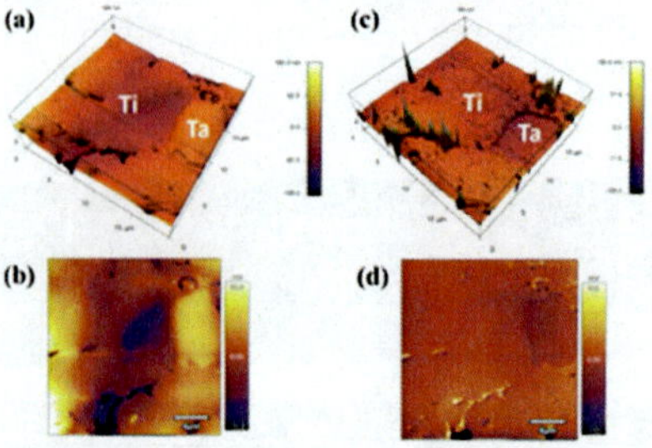

Fig. 3. SKPFM measurements on moderate-porous Ti-Ta composite (polished surface). (a), (b) 3D and 2D surface morphologies; (c), (d) 3D and 2D surface potential distributions.

4 Conclusions

This study explores the influences of porosity on the *in-vitro* cell responses of cold sprayed Ti-Ta composites. Cell growth rates were significantly enhanced on cold sprayed moderate-porous Ti-Ta composite compared to the low-porous ones especially during the initial few hours' cell incubation period. It is evident that cold spray additive manufacturing would be utilized to surface topographical modifications with benefits to enhance the osseointegration for biomedical applications.

References

1. Rabel, K., et al.: Controlling osteoblast morphology and proliferation via surface micro-topographies of implant biomaterials. Sci. Rep. **10**, 1–14 (2020)
2. Bartolomeu, F., Dourado, N., Pereira, F., Alves, N., Miranda, G., Silva, F.: Additive manufactured porous biomaterials targeting orthopedic implants: a suitable combination of mechanical, physical and topological properties, Materials Sci. Eng.: C **107** 110342 (2020)
3. Danilevicius, P., Georgiadi, L., Pateman, C.J., Claeyssens, F., Chatzinikolaidou, M., Farsari, M.: The effect of porosity on cell ingrowth into accurately defined, laser-made, polylactide-based 3D scaffolds. Appl. Surf. Sci. **336**, 2–10 (2015)
4. Wauthle, R., et al.: Revival of pure titanium for dynamically loaded porous implants using additive manufacturing. Mater. Sci. Eng. C **54**, 94–100 (2015)
5. Zeng, G., Zahiri, S., Gulizia, S., Chen, Y., Chen, X.-B., Cole, I.: A comparative study of cell growth on a cold sprayed Ti–Ta composite. J. Alloys Comp. **826**, 154014 (2020)
6. Li, W., Cao, C., Yin, S.: Solid-state cold spraying of Ti and its alloys: a literature review. Prog. Materials Sci. **110**, 100633 (2020)

Modification of Cold Sprayed CoCrMo Alloy Coatings via Laser Shock Peening

Wen Sun[1,2](✉), Niroj Maharjan[3], Kaiqiang Wu[4], Adrian Wei-Yee Tan[5], Renzhong Huang[1], Yingchun Xie[1], Haiming Lan[1], Xin Chu[1], and Erjia Liu[4]

[1] National Engineering Laboratory for Modern Materials Surface Engineering Technology, Institute of New Materials Guangdong Academy of Sciences, Guangzhou 510650, China
SUNW0013@e.ntu.edu.sg

[2] School of Materials Science and Engineering, South China University of Technology, Guangzhou 510641, China

[3] Advanced Remanufacturing and Technology Centre, CleanTech Loop, Singapore 637143, Singapore

[4] School of Mechanical and Aerospace Engineering, Nanyang Technological University, 50 Nanyang Avenue, Singapore 639798, Singapore

[5] School of Engineering, University of Southampton Malaysia, 79200 Iskandar Puteri, Johor, Malaysia

Abstract. In this study, CoCrMo alloy coatings were fabricated by high pressure cold spray deposition. Laser shock peening with and without an ablative layer was applied to the coatings. The microstructure characteristics and mechanical properties of the both as-sprayed and laser shock peened CoCrMo coatings were systematically investigated. Laser shock peening without a protective overlay resulted in both mechanical and thermal loading of the coatings, leading to melting and solidification of the outermost surfaces of the coatings along with deformation. For comparison, laser shock peening with a polyvinyl tape as sacrificial layer was also applied to the as-sprayed CoCrMo coatings. It was found that this sacrificial layer could protect the samples from thermal effects and introduce a deep compressive residual stress in the coatings, which resulted in cracks in the coating. Those cracks in the coatings appeared to be continuous and parallel to the coating/substrate interface due to the strong shock wave imposed during the laser peening process.

Keywords: High pressure cold spray · CoCrMo alloy coating · Laser shock peening · Microstructure · Mechanical properties

1 Introduction

Cold spray is an emerging technology that can be used for coating deposition, metal surface repair and additive manufacturing, which has attracted more and more attention in recent years [1]. The fundamental working principle of cold spray technology is that micro-sized metal particles are accelerated by a high-pressure gas to a supersonic speed via a convergent-divergent nozzle. With high kinetic energies, the metal particles deposit

Y. Wei and S. Chng (Eds.): INCASE 2021, LNME, pp. 185–188, 2022.
https://doi.org/10.1007/978-981-16-5763-4_41

on the substrate surface and form a dense coating of a high adhesive strength with a low temperature input. Cold spray deposition is applicable to nearly unlimited materials, such as pure metals, metal alloys, polymers, and hybrid materials [2, 3]. Thus, cold spray technology holds great potential as a next-generation large-scale and low-cost coating/additive manufacturing technology in industry [4].

Micropores and microcracks within the as-sprayed coating are inevitable due to the bonding nature of cold spray particles, which can lead to the degradation of mechanical properties of the coating and its bonding to the substrate [5, 6]. Sun et al. [7] reviewed the effects of various post-process treatments on the microstructures and properties of the cold sprayed coatings. However, each method has its benefits and limitations, and further post-process treatment methods need to be explored.

Laser shock peening (LSP) is an innovative contact-free surface modification technique, which is employed to improve the material performances, such as fatigue life, anti-corrosion, and wear resistance. The LSP process utilizes short and high energy laser pulses to generate a high-pressure plasma that can impart surface compressive residual stress, increase surface hardness and refine grain of metals [8]. This study aims to evaluate the effects of LSP on the microstructures and mechanical properties of the cold sprayed CoCrMo coatings.

2 Materials and Methods

The feedstock power material used in this study was F75 CoCrMo (TLS Technik GmbH & Co, Germany) with average particle size of 25 μm. The CoCrMo particles are spherical. The substrates used in this study were Inconel 718. A high-pressure cold spray system was used for the deposition of the CoCrMo coatings on the IN718 substrates. N_2 was used as the propelling gas.

Before LSP, the as-sprayed CoCrMo coatings were cut into 50 mm $\times$ 20 mm $\times$ 10 mm and subsequently cleaned in deionized water. All the specimens were degreased by ultrasound in ethanol. A Nd:YAG laser system with a 1.064 μm wavelength, 10 J pulse energy, 18 ns laser pulse width and 4 Hz repetition rate was used to peen the samples in presence of water confinement. Two sets of samples were made: with a protective coating and without a protective coating. A black vinyl tape was used as a protective coating.

3 Results and Discussions

Figure 1 shows the optical micrographs of the cross-sectional CoCrMo coatings under different conditions. It can be observed that the as-sprayed CoCrMo coating consists of a very low porosity level, which is around 0.7%, as shown in Fig. 1(a). The irregular shape micro-pores are uniformly distributed within the coating. The interface between the coating and substrate is intimate without signs of delamination, indicating a strong bonding. Besides, there is no obvious cracks within the as-sprayed coating. However, after LSP treatments, some changes for the coating microstructures occur. Figure 1(b) shows the microstructures after LSP with an ablative layer. It can be seen that a continuous crack exists within the coating, which is almost parallel to the coating/substrate interface due to the very high amplitude pressure induced by LSP. Figure 1(c) shows the microstructures after LSP without an ablative layer. It can be seen that no obvious crack exists within the coating because of the combined mechanical and thermal effects of the LSP.

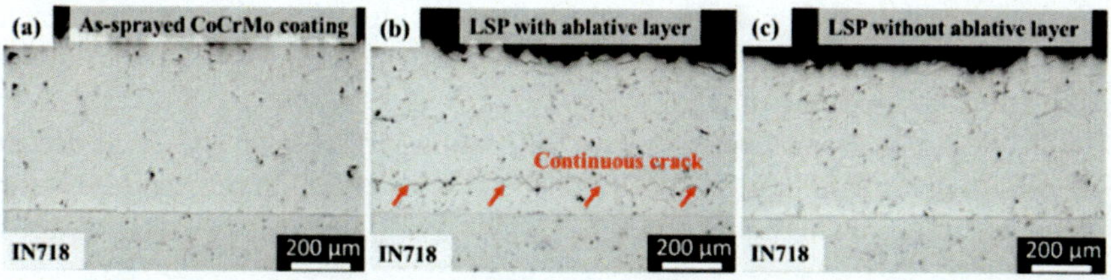

Fig. 1. Optical micrographs showing cross-sectional microstructures of cold sprayed CoCrMo coatings on IN718 substrates: (a) as-sprayed, (b) LSP with an ablative layer and (c) LSP without an ablative layer.

Figure 2 shows the micro-hardness values of CoCrMo coatings and IN718 substrates before and after LSP treatment. It can be seen that laser shock peening (with or without an ablative layer) could improve the micro-hardness of the coating surface because of plastic deformation, high density dislocations and fine grains induced on the material surface by high-pressure shock waves. With the increase of depth, laser shock peening has less effect on the coating micro-hardness because of the reduced strength of shock waves.

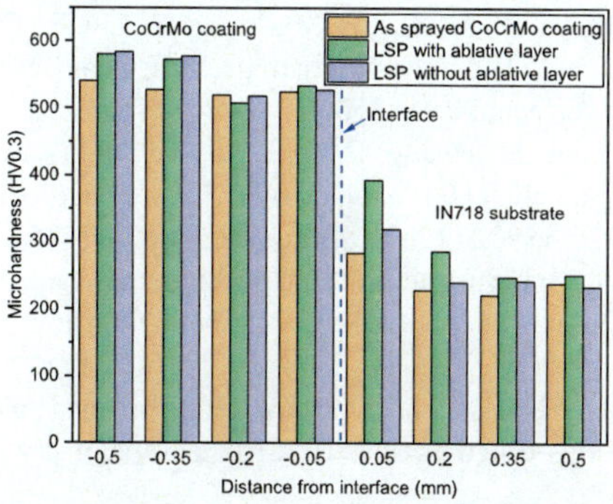

Fig. 2. Microhardnesses of CoCrMo coating and IN718 substrate before and after LSP.

4 Conclusions

In this study, CoCrMo alloy coatings were fabricated by high pressure cold spray deposition. Laser shock peening was applied to the coatings. The microstructure characteristics and mechanical properties of the CoCrMo coatings were systematically investigated. The main conclusions are drawn as below:

(1) For the laser shock peening without an ablative layer, the coating surfaces showed presence of a non-uniform melting layer that increased the hardness and roughness of the surfaces.
(2) For the laser shock peening with a polyvinyl tape as sacrificial layer, it was found that this sacrificial layer could protect the samples from thermal effects, which resulted in cracks in the coating. Those cracks in the coatings appeared to be continuous and parallel to the coating/substrate interface due to the strong shock wave imposed during the laser peening process.
(3) Laser shock processing could improve the coating surface micro-hardness because of plastic deformation, high density dislocations and fine grains induced on the material surface by high-pressure shock waves.

Acknowledgements. This work was supported by Guangdong Province Key Area Research Grant: 2020B0101330001. Portions of the work were performed as part of collaborative research project (FYP B180–2019) between Advanced Remanufacturing Technology Centre (ARTC) and Nanyang Technological University (NTU), Singapore.

References

1. Tan, A.W.-Y., et al.: Microstructure, mechanical and tribological properties of cold sprayed Ti6Al4V–CoCr composite coatings. Compos. Part B: Eng. **202**, (2020)
2. Sun, W., et al.: Effect of substrate surface condition on fatigue behavior of cold sprayed Ti6Al4V coatings. Surf. Coat. Technol. **320**, 452–457 (2017)
3. Sun, W., Tan, A.W.-Y., Bhowmik, A., Xue, F., Marinescu, I., Liu, E.: Evaluation of cold sprayed graphene nanoplates–Inconel 718 composite coatings. Surf. Coat. Technol. **378**, 125065 (2019). https://doi.org/10.1016/j.surfcoat.2019.125065
4. Wei, Y.-K., Luo, X.-T., Chu, X., Huang, G.-S., Li, C.-J.: Solid-state additive manufacturing high performance aluminum alloy 6061 enabled by an in-situ micro-forging assisted cold spray. Materials Sci. Eng.: A **776**, 139024 (2020). https://doi.org/10.1016/j.msea.2020.139024
5. Sun, W., et al.: Improving microstructural and mechanical characteristics of cold-sprayed Inconel 718 deposits via local induction heat treatment. J. Alloy. Compd. **797**, 1268–1279 (2019)
6. Khun, N.W., et al.: Effects of Nd:YAG laser surface treatment on tribological properties of cold-sprayed Ti-6Al-4V coatings tested against 100Cr6 steel under dry condition. Tribol. Trans. **62**(3), 391–402 (2019)
7. Sun, W., Tan, A.W.-Y., Kaiqiang, W., Yin, S., Yang, X., Marinescu, I., Liu, E.: Post-process treatments on supersonic cold sprayed coatings: a review. Coatings **10**(2), 123 (2020). https://doi.org/10.3390/coatings10020123
8. Maharjan, N., et al.: Fatigue performance of laser shock peened Ti6Al4V and Al6061-T6 alloys. Fatigue Fract. Eng. Mater. Struct. **44**(3), 733–747 (2021)

Feedstock Characteristics Effect on Cold Sprayability of Tantalum Coatings

Xin Chu[1,2](✉), Huseyin Adyin[3], Renzhong Huang[1], Yingchun Xie[1], Wen Sun[1], Haiming Lan[1], and Steve Yue[3]

[1] National Engineering Laboratory for Modern Materials Surface Engineering Technology, Institute of New Materials, Guangdong Academy of Sciences, Guangzhou 510651, China
chuxin@gdinm.com

[2] School of Materials Science and Engineering, South China University of Technology, Guangzhou 510641, China

[3] Department of Mining and Materials Engineering, McGill University, Montreal, QC H3A0C5, Canada

Abstract. In this study, two spherical (P1: hard, P2: soft) tantalum powders as well as their binary mixtures (P1 + 20 wt.% P2) were deposited using high pressure cold spray to investigate the powder characteristics effect. The coating microstructure and cold sprayability metrics (porosity, deposition efficiency (DE), cohesion strength) were evaluated. Results show that i) soft tantalum powder (P2) exhibits slightly higher deposition efficiency (DE) and much higher cohesion strength than the hard tantalum powder (P1); ii) mixing tantalum powders with different hardnesses (P1 + P2) can lead to improved DE and strength compared with single component P1 and P2 powders.

Keywords: Cold spray · Tantalum · Powder characteristics · Mixing powders · Cold sprayability

1 Introduction

Tantalum (Ta) is a heavy refractory metal (density: 16.65 g/cm^3; melting point: 2996 °C) and possesses many important properties such as extreme corrosion resistance and biocompatibility [1]. It is also an expensive 'exotic' metal and is thus best used as coatings on less expensive alloy substrates. Cold spray is a promising coating and solid state additive manufacturing technology which utilizes a high pressure gas to accelerate the solid particles and impact on the substrate to form the deposit [2]. It is rapid and its working temperature is well below the melting point, thus avoiding heat related issues such as oxidation and thermal stresses [2]. Limited studies on cold spray of tantalum show it can be successfully deposited with a deposition efficiency (DE) of 50–100% [3–5]. Although tantalum is a body centered cubic (BCC) metal, its high level of cold sprayability can be due partly to its relatively soft nature [4]. However, the mechanical properties (e.g. strength, ductility) of bulk Ta are also reported to be strongly dependent on its impurity levels, e.g. oxygen (interstitial element) [6]. For an increase from 300 to 1200 ppm

Y. Wei and S. Chng (Eds.): INCASE 2021, LNME, pp. 189–192, 2022.
https://doi.org/10.1007/978-981-16-5763-4_42

oxygen, its tensile strength increases from 300 to 900 MPa and ductility (reduction in area) decreases from 80% to 0% [6]. It is also noted that a good cold sprayability of a feedstock powder (e.g. high DE) does not necessarily guarantee satisfactory mechanical or functional properties of the coating.

In this study, two tantalum powders (hard-P1, soft-P2) as well as their binary mixture (P1 + 20wt.%P2) were cold spray deposited. The coating microstructure (cross-sections, top surfaces) was characterized. The cold sprayability (porosity, DE, cohesion strength) of different coatings were evaluated. Based on the results, the effects of powder characteristics and mixing powders on cold sprayability of tantalum coatings were discussed.

2 Materials and Methods

The feedstock powders used in this study are high purity (>99.9%) tantalum powders supplied by Tekna Plasma System Inc. (Sherbrooke, Canada), as shown in Fig. 1. P1 is a commercial available one and P2 is a specifically prepared one to reduce oxygen content by using spheroidized HC Starck feedstock. P1 is slightly smaller in size and the re-melted P2 has slightly lower hardness compared with P1 and P2, which is consistent with the oxygen content. One their binary mixture was also prepared by using P1 + 20wt.%P2 (P1 + P2). The cold spray was conducted using a commercial high-pressure cold spray system (Plasma Giken PCS-800, Japan) using a gas pressure of 4 MPa and gas temperature of 800 °C. The propellant gas was nitrogen.

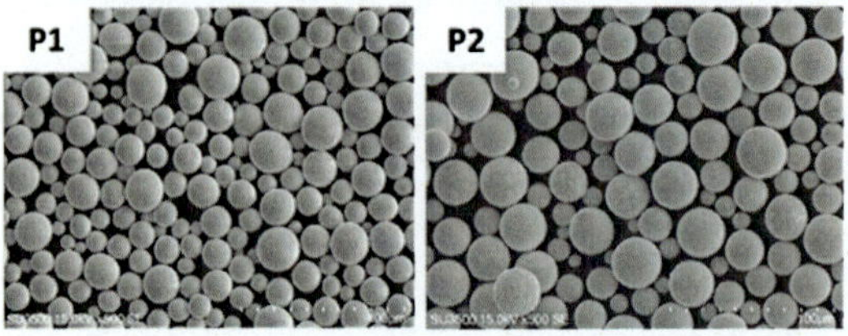

Powder	Mean size, µm	Hardness, $HV_{0.01}$	Oxygen, ppm
P1	19	167	405
P2	24	145	259

Fig. 1. SEM images and characteristics of the feedstock powders.

3 Results and Discussions

Figure 2 shows the cross-sectional optical micrographs of the deposited coatings. It can be seen that all three coatings (single component and mixed) exhibit very low porosity levels and are free of cracks. The porosity levels measured are all below 0.5%. This can be ascribed to the relatively soft nature of the tantalum powders.

Figure 3 shows the cold sprayability metrics of the as-deposited coatings. The predicted values of mixed coatings were calculated based on the rule of mixtures ($w_{mix} = f_1 * w_1 + f_2 * w_2$). Figure 3(a) shows that the deposition efficiency (DE) and P2 has slightly higher DE than P1 (65% vs 60%). Mixing P1 and P2 (P1 + P2) resulted in a much higher DE than P1 and P2 alone. Figure 3(b) shows the ultimate tensile strength

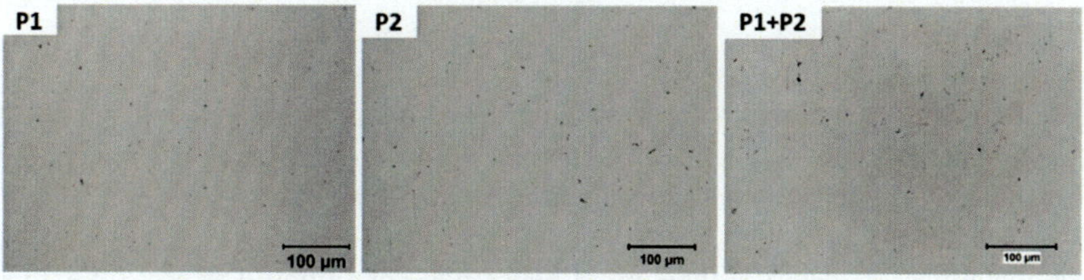

Fig. 2. Cross-sectional microstructure of different coatings.

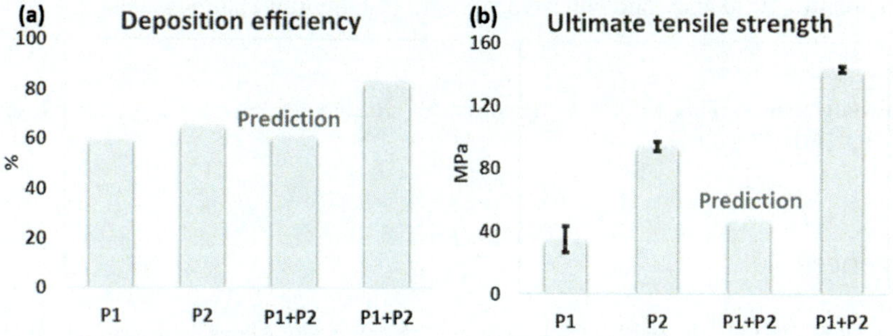

Fig. 3. Cold sprayability of the coatings. a) DE and (b) ultimate tensile strength.

(UTS) of the as-deposited coatings. P2 has a much higher strength than P1 mainly due to its softer nature while P1 + P2 generates the highest strength among all the coatings.

Figure 4 shows the SEM top surfaces of the deposited coatings. For P1 coating, only flattened particles and craters (left by rebounded particles) can be observed and particle surfaces are all smooth. For P1 + P2 mixture, the particles are more deeply embedded into compared with single component P1. Also, some materials are observed to exhibit outward metal jets. In cold spray, the metal jets are associated with metallurgical bonding and thus better cold sprayability [3–5]. It only occurs in P1 + P2 where a hardness difference among feedstocks is created. A simplified schematic is thus created accordingly.

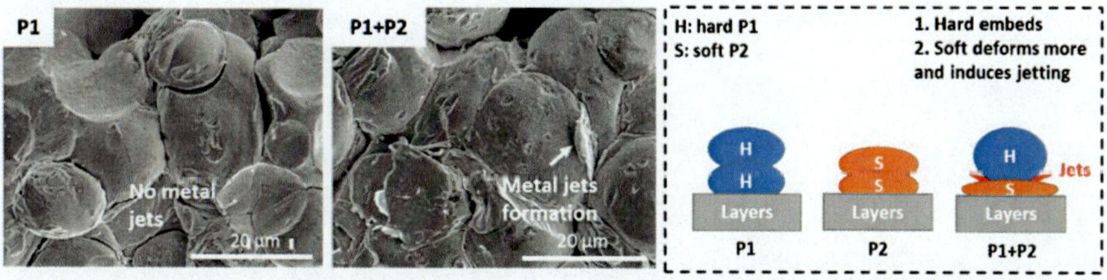

Fig. 4. Top surface morphology of different coatings and a proposed deposition schematic.

4 Conclusions

In this study, three tantalum coatings were produced by high pressure cold spray. The coating microstructure was characterized and the cold sprayability of different coatings were evaluated. The main conclusions are drawn as follows:

(1) All coatings are dense at the current deposition conditions. The soft P2 exhibits slightly higher deposition efficiency and much higher strength than the hard P1.
(2) Creating a hardness difference among feedstocks (P1 + P2) can lead to improved deposition efficiency and cohesive strength of tantalum coatings.

Acknowledgements. This work was supported by Guangdong Province Key Area Research Grant: 2020B0101330001.

References

1. Committee, A.I.H.: Properties and Selection: Nonferrous Alloys and Special- Purpose Materials. ASM International (1990)
2. Karthikeyan, J.: The advantages and disadvantages of the cold spray coating process. In: The Cold Spray Materials Deposition Process, pp. 62–71. Elsevier (2007). https://doi.org/10.1533/9781845693787.1.62
3. Kim, J., Bae, G., Lee, C.: Characteristics of kinetic sprayed Ta in terms of the deposition behavior, microstructural evolution and mechanical properties: Effect of strain-rate-dependent response of Ta at high temperature. Mater. Charact. **141**, 49–58 (2018)
4. Koivuluoto, H., Näkki, J., Vuoristo, P.: Corrosion properties of cold-sprayed tantalum coatings. J. Therm. Spray Technol. **18**(1), 75–82 (2008)
5. Van Steenkiste, T., Gorkiewicz, D.W.: Analysis of tantalum coatings produced by the kinetic spray process. J. Therm. Spray Technol. **13**(2), 265–273 (2004)
6. Martienssen, W., Warlimont, H. (eds.): Springer handbook of condensed matter and materials data. Springer Berlin Heidelberg (2005). https://doi.org/10.1007/3-540-30437-1

NoVOC Water-Borne Coatings Through Utilization of the Nano-confinement Effect

Praveen Thoniyot(✉), Ritwik Panigrahi, and Alexander van Herk(✉)

Institute of Chemical and Engineering Sciences, Singapore, Singapore
{thoniyotp,alexanderherk}@ices.a-star.edu.sg

Abstract. Water-borne coatings are environmentally friendly compared solvent born coatings where polymer particles (the binder) and other formulation ingredients are dispersed in water. However, most water-borne coatings contain 5–10% of organic molecules like plasticizers and coalescing agents. These organic molecules evaporate from the drying coating posing an environmental burden. Many approaches have been investigated, but there is still no good working alternative for the plasticizers. The coalescing agents temporarily lower the *Tg* of the binder to facilitate particle deformation, after evaporation the glass transition temperature is restored to the original *Tg* again. A versatile alternative is to use lower *Tg* binders and create a "jump" in *Tg* during film formation. Our wok uses this nano-confinement effect, which creates a jump in *Tg* when the polymer chains are coming in close contact with inorganic nanoparticles surfaces, get immobilized. This close contact happens in the particle deformation stage, automatically leading to the correct timing of the two processes, particle deformation and increase of *Tg*. Another advantage of this approach is that no new chemistries need to be introduced. However it turns out that this approach only properly functions in a narrow window of nanoparticle loading.

Keywords: Water-borne coatings · NoVOC coatings · Nanoparticles · Inorganic surface-polymer interactions · Glass transition temperature change

1 Introduction

Confinement effects due to the presence of hard nanoparticles in a polymer matrix have a profound effect on polymer T_g [1]. So far the nano-confinement effect has been studied from casting films out of a polymer solution in the presence of inorganic nanoparticles [1]. The interactions between the polymer chain segments and/or the end groups of the chain play an important role, leading to either an increase (for interaction) or a decrease in (repulsive forces) *Tg*. We demonstrate the effect of end groups in determining the *Tg* via a simple equation to work out the concentration of nanoparticles for maximum T_g in a composite for a given polymer. Using polystyrene-silica system with various end groups and polyvinyl alcohol-silica nanocomposite with interacting groups present throughout the polymer chain, we demonstrate that loading of nanoparticles for maximum T_g is governed mostly by interactions of chain ends with the nanoparticle surface.

Y. Wei and S. Chng (Eds.): INCASE 2021, LNME, pp. 193–196, 2022.
https://doi.org/10.1007/978-981-16-5763-4_43

2 Methods

For the polyvinyl alcohol system with silica we used aqueous solution of PVA of different molecular weights and cast films by drying For the polystyrene-silica combinations we used solutions in THF to cast the film [2]. We also studied whether similar *Tg* jumps can also be observed in latex systems. For the latex systems we used composite of latex and silica and analyzed the *Tg* using Dynamic Mechanical Analysis and Differential Scanning Calorimetry. Finally, a recipe for water-borne coatings that introduces nano-silica particles in the formulation according to the experimental composition that gave maximum *Tg* was prepared. The corresponding coatings films were tested for gloss, scratch resistance and other typical paint film tests.

3 Results and Discussion

We observed the *Tg* jump phenomena due to the nano-confinement effect in PVA-silica, polystyrene-silica and latex-silica systems. We studied water based PVA system in depth under controlled conditions to establish the relationship between nanoparticle size and polymer molecular weight. A typical result when used silica nanoparticle (Fig. 1).

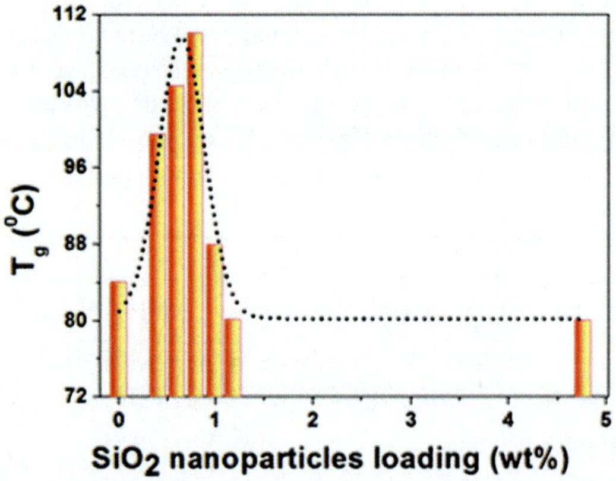

Fig. 1. Effect of silica loading on *Tg* (deduced from dynamic mechanical analysis) in PVA-silica composite prepared in water. Composites prepared in water with a solution of PVA (degree of polymerization of 2000) and suspension of silica nanoparticles (average diameter -24 nm).

For the first time the nano-confinement effect was also shown to occur in a latex system by studying the composites formed by latex and varying amount of silica nanoparticles A typical example of *Tg* effect in latex-silica composite system is shown in Fig. 2.

The *Tg* increase in the latex system would mean that during the film formation process, at the stage where the particle deformation and polymer diffusions happens, the polymer chains are starting to interact with the surface of the inorganic particles through hydrogen bonding between hydroxyl groups on the surface and acid groups on the latex.

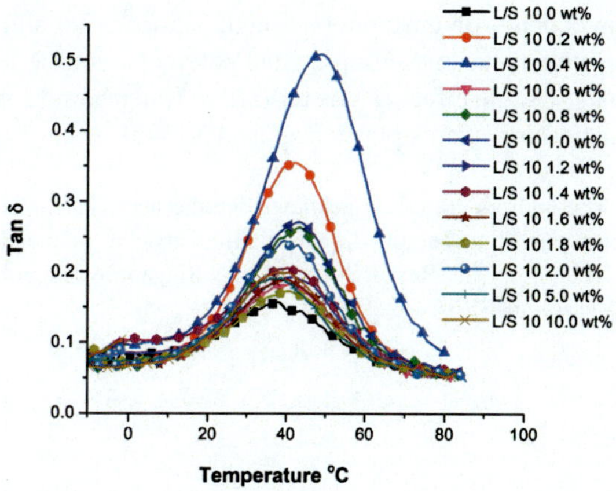

Fig. 2. Effect of *Tg* (deduced from dynamic mechanical analysis) on silica loading in Sty, HEA, MMA and MAA latex-silica composite prepared in water (silica nanoparticles average diameter -10 nm).

This effect was evident in a non-tacky film obtained when we used a low *Tg* latex (15°C) in the final formulation.

Surprisingly all the *Tg* effects showed an optimum at a certain loading of the nanoparticles (Figs. 1 and 2). This optimum loading corresponding to the highest *Tg* depends on the particle size of the nanoparticles as well as the molecular weight of the polymers and their end groups.

We were able to derive an equation that predicts the location of this optimal loading.

For the latex systems, the resulting coatings performed like their regular counterparts, with the difference that no coalescing agents/plasticizers were added and instead a lower *Tg* latex was utilized (with the regular co-stabilizing acidic comonomers) and a low loading of commercial inorganic nanoparticles. The crucial factor is to select the optimum loading of these inorganic nanoparticles which can be predicted via our model but would otherwise be relatively difficult to establish experimentally.

The question might be asked why this effect does not play a role in regular coating formulations with an abundance of inorganic (pigment and filler) particles. The answer lies in the size of these particles, representing a much smaller surface area than the typical nanoparticles used in these studies, being in the 10–50 nm range whereas the pigment particles are approximately 10 times bigger and the fillers are at least 100 times bigger. To test whether this *Tg* effect finally translates to the preparation of SOC free water-based paint formulations we prepared coating formulations with low *Tg* latex and bench marked its performance with commercial formulations. We did not see any notable difference in performance in terms of color, gloss, tackiness or visual defects and the coating performed well in a one-year outdoor testing.

4 Conclusion

The nanoconfinement effect can be observed in films cast from solutions. We have shown that there is a strong influence on the loading of nanoparticles where this *Tg* jump is highest. The location depends on the combination of nanoparticles and polymer and the molecular weight as well as the end groups of the polymers. For the first time we were also able to show that the same effect can be utilized in water-borne coatings, leading to the elimination of coalescing agents/plasticizers and NoVOC water-borne coatings.

Acknowledgement. The authors gratefully acknowledge the National Supercomputing Centre (NSCC), Singapore, for providing computational facilities and the financial support from the Agency for Science, Technology and Research (A*STAR), Singapore, Environmentally Friendly Specialty Products Programme, SERC152800043.

References

1. Rittigstein, P., Priestley, R.D., Broadbelt, L.J., Torkelson, J.M.: Model polymer nanocomposites provide an understanding of confinement effects in real nanocomposites. Nat. Mater. **2007**(6), 278–282 (2007)
2. Panigrahi, R., et al.: Elucidating the role of interfacial hydrogen bonds on glass transition temperature change in a Poly (Vinyl Alcohol)/SiO2 polymer-nanocomposite by noncovalent interaction characterization and atomistic molecular dynamics simulation. Macromol. Rapid Commun. **41**(21), 2000240 (2020). https://doi.org/10.1002/marc.202000240

Modelling and Simulation

Calibration of a Continuum-Based Granular Flow Dynamics Model for the Numerical Simulation of the Stream Finishing Process

Stephen Wan[1](✉), Shengwei Ma[1], Cary Turangan[1], Keni Chih-Hua Wu[1], Sho Itoh[2], Wei Shin Cheng[3], Kai Liang Tan[3], and Jeremy Ho[3]

[1] Institute of High Performance Computing, 1 Fusionopolis Way, #16–16 Connexis, Singapore, Singapore
{wansym,masw,cary,keni-wu}@ihpc.a-star.edu.sg

[2] Graduate School of Engineering, Faculty of Engineering, Chiba University, 1-33 Yayoi-cho, Inage-ku, Chiba-shi, Chiba 263-8522, Japan
ittosyo@chiba-u.jp

[3] Advanced Remanufacturing and Technology Centre, 3 Cleantech Loop, #01/01, CleanTech 2, Singapore, Singapore
{chengws,tan_kai_liang,jeremy-ho}@artc.a-star.edu.sg

Abstract. The present study spearheads an ongoing effort aimed at developing a computationally inexpensive and reasonably accurate numerical simulation model for the prediction of the material removal distribution over the surfaces of additively manufactured parts polished by the stream finishing process. To this end, the parameters of a continuum-based granular flow dynamics model are calibrated against experimental data obtained by interrogating the media flow field, in a typical stream finishing drum, with a purpose-designed probe. Interestingly, it was observed that, including a shear stress and pressure dependent wall-slip wall model yielded a better comparison with the experimental data; whilst, the prognosis is that a free-slip condition appears adequate as a first approximation.

Keywords: Stream finishing · Numerical simulation · Continuum-based granular flow dynamics model

1 Introduction

The present study represents the initial stage of an ongoing effort, which is driven by a need to numerically predict the material removal distribution over the surface of additive manufactured parts processed by stream finishing in order to reduce the number of trial-and-error experimental iterations.

Stream finishing is a mass finishing process in which a part is immersed in a rotating drum filled with abrasive media. The process is capable of very aggressive material removal rates, compared with other mass finishing techniques; and may be considered as the reverse of the drag finishing process, in which the drum is held stationary instead, and the part, forced through the abrasive media.

Y. Wei and S. Chng (Eds.): INCASE 2021, LNME, pp. 199–202, 2022.
https://doi.org/10.1007/978-981-16-5763-4_44

Perhaps the earliest work on stream finishing reported in the open literature belongs to Schulze et al. [1], who studied the effect of processing time on the surface micro hardness, residual stresses and roughness of AISI 4140 steel specimens. This study was later extended by Karacas et al. [2] to include to the effect of drum rotational speed. Zanger et al. [3] undertook a combined experimental and numerical analysis for a deeper understanding of the relationships between various process parameters and normal forces, material removal and surface topography for AISI 4140 specimens. They modelled the stream finishing process using the discrete element method (DEM); whereas, a preliminary attempt at a continuum – based approach was made by Turangan et al. [4] and tested against experimental data from Ito et al. [5].

2 Methods

2.1 Continuum – Based Granular Flow Dynamics Modelling Approach

The flow of media in the rotating drum is modelled by a continuum, the rheology of which, in simple terms, can be described by an inelastic non-Newtonian viscosity model. Specifically, we have chosen the model explicated in [6] because of its simplicity. The viscosity, η is given by:

$$\eta = \frac{\mu(I)P}{|\gamma|} \tag{1}$$

in which, P is the isotropic pressure; $|\gamma| = \sqrt{\frac{\gamma_{ij}\gamma_{ij}}{2}}$ is the strain – rate magnitude; and $\mu(I)$ is a friction coefficient given by:

$$\mu(I) = \mu_1 + \frac{\mu_2 - \mu_1}{I_0/I + 1}(2) \tag{2}$$

where $I = |\gamma| d / \sqrt{P/\rho_p}$ is termed the inertial number; d and ρ_p are the media particle size and density respectively; and, in the context of this study, μ_1, μ_2 and I_0 are the rheological parameters to be calibrated against experimental data.

2.2 Numerical simulation set-up

Numerical simulations were carried out using the computational fluid dynamics (CFD) software package, ANSYS CFX version 17.2, with the viscosity function, given by Eqs. (1) and (2), coded in accordingly. The boundary conditions of the computational domain are shown in Fig. 1. The wall slip boundary condition imposed on the surfaces on the probe is given by a wall shear – stress and pressure load dependent slip model which gives the wall slip velocity, u_w as [7]:

$$u_w = u_s \left(\frac{\tau_w}{\tau_n}\right)^m \exp\left(-\frac{BP}{\tau_n}\right) \tag{3}$$

in which, τ_w is the wall shear stress; P is wall pressure; τ_n is a scaling wall shear – stress; and u_s, m and B are the model parameters to be calibrated against experimental data.

2.3 Experimental Set - up

Experimental data for calibrating the rheology and wall slip model parameters were acquired by measuring slip velocities across and pressures acting on a purpose built probe at various radial arm distance, r, media depth, z and glancing angles, θ_z, as indicated in Fig. 2, for drum rotational speed of 30 RPM. Tetrahedral shaped ceramic DBS 4X4 media ($d = 4\text{mm}$, $\rho_p = 2083\text{kgm}^{-3}$) were used in the experiments.

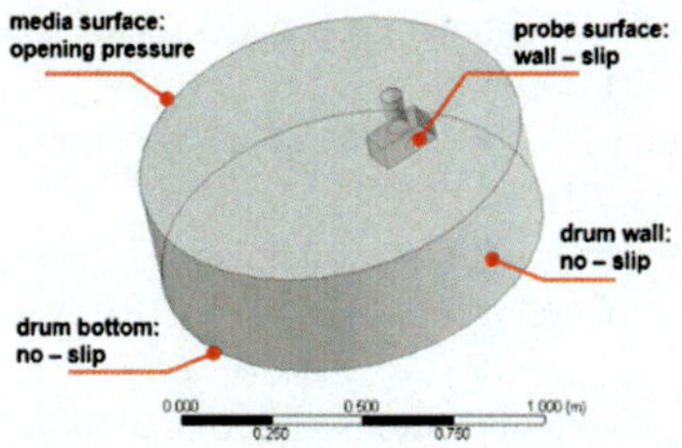

Fig. 1. Boundary conditions imposed on computational flow domain.

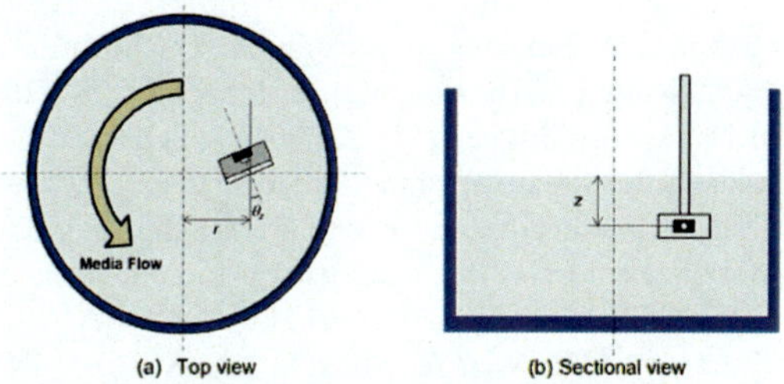

Fig. 2. (a) Top and (b) side view of experimental setup showing probe and direction of drum rotation.

3 Results and Discussion

A preliminary attempt at calibrating the parameter values for the rheology model (Eq. (2)) gave: $\mu_1 = 0.3$, $\mu_2 = 0.65$ and $I_0 = 0.3$; and those for the wall-slip model (Eq. 3): $u_s = 0.3\text{ms}^{-1}$, $m = 1$ and $B = 0.12$. With these values, representative comparisons of the experimental and numerical vertical and horizontal wall slip velocities (v_z and v_h) and pressure load, F with respect with glancing angle, θ_z are shown in Figs. 3 and 4, respectively.

Both the magnitudes of vertical and horizontal wall slip velocities (v_z and v_h) tend to be over-predicted, while the pressure load, F tends to be under-predicted at higher glancing angles, θ_z by the simulation model.

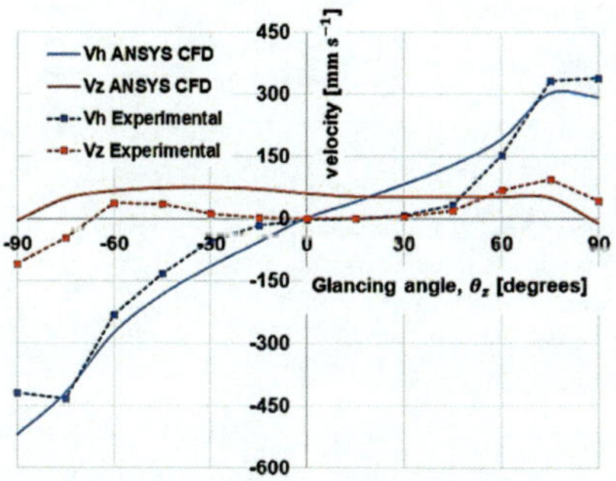

Fig. 3. Comparison of vertical and horizontal slip velocities for a media depth, $z = 200\text{mm}$.

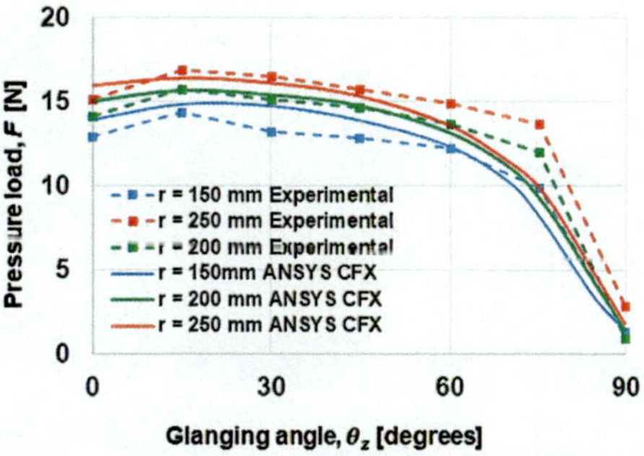

Fig. 4. Comparison of pressure load, F, for a media depth, $z = 150\text{mm}$ at various radial arm distance, r.

4 Conclusions

While the results of the present study are promising, discrepancies between simulation and experimental results warrant a closer look at model limitations as well as measurement uncertainties inherent in the experimental methodology.

References

1. Schulze, V., Gibmeier, J., Kacaras, A.: Qualification of the stream finishing process for surface modification. CIRP Ann. Manuf. Tech. **66**, 523–526 (2017)
2. Kacaras, A., Gibmeier, J., Zanger, F., Schulze, V.: Influence of rotational speed on surface states after stream finishing. Procedia CIRP **71**, 221–226 (2018)
3. Zanger, F., Kacaras, A., Neuenfeldt, P., Schulze, V.: Optimization of the stream finishing process for mechanical surface treatment by numerical and experimental process analysis. CIRP Ann. Manuf. Tech. **68**, 373–376 (2019)
4. Turangan, C.K., Wan Yee Ming, S., Itoh, S., Ho, J.: Development of rheology and computational flow model for robotized external finishing on additively manufactured components. In: Itoh, S., Shukla, S. (eds.) INCASE 2019. LNME, pp. 234–243. Springer, Singapore (2020). https://doi.org/10.1007/978-981-15-0054-1_24
5. Itoh, S., Ho, J., Turangan, C., Wan, S.: In situ measurement of granular pressure and velocity on component surfaces in stream finishing. In: Itoh, S., Shukla, S. (eds.) INCASE 2019. LNME, pp. 226–233. Springer, Singapore (2020). https://doi.org/10.1007/978-981-15-0054-1_23
6. Forterre, Y., Pouliquen, O.: Flows of dense granular media. Annu. Rev. Fluid Mech. **40**, 1–24 (2008)
7. Person, T.J., Denn, M.M.: The effect of die materials and pressure-dependent slip on the sextrusion of linear low density polyethylene. J. Rheol. **41**, 249 (1997)

FDTD Simulation of the Spectral Characteristics of the Secondary Array Nanostructures on the Material Surface

Lu Meirong, Liu Shibing, Wang Bing(✉), and Song Haiying(✉)

Strong-Field and Ultrafast Photonics Lab, Key Laboratory of Trans-Scale Laser Manufacturing Technology, Ministry of Education, Faculty of Materials and Manufacturing, Beijing University of Technology, Beijing 100124, China
{wangbing,hysong}@bjut.edu.cn

Abstract. In this work, we use a finite-difference time-domain (FDTD) method to study the spectral changes of the convex array structure on the surface of titanium (Ti) from visible to near infrared. The influence of different structural parameters of the convex structure on the reflection spectrum is studied. It is believed that the selective absorption of spectrum is due to the surface plasmon effect. For the specific structural parameters, the absorption can reach 100% at a specific wavelength. On the other hand, spectrum characteristics of the surface when the convex array structure is covered with ZnO nanorods are studied. The results show that when the surface covered with ZnO nanorods, the number of absorption peaks increases and the absorptivity increases as a whole. The influence of different structural parameters of ZnO nanorods on the reflection spectrum is discussed.

Keywords: FDTD method · The convex array structure · ZnO nanorods · Reflection spectrum

1 Introduction

Studies have shown that surface microstructures with a feature size similar to the incident wavelength have a significant regulatory effect on the spectral characteristics, and various types of microstructures have different effects on the control of their spectral characteristics [1–12]. Based on the surface plasmon properties, for metals, there are usually a lot of free electrons on the surface, so as long as the appropriate microstructure is constructed on the surface, it is easy to realize the spectral control in different wavelength bands. Such as, the absorption and radiation characteristics of metamaterial nanostructures made of two-dimensional tungsten gratings were numerically studied, the spectral absorptivity of the designed absorber is higher than 0.95 in the wavelength range of 0.6 μm to 1.8 μm, while the mid-infrared spectral absorptivity is lower than 0.04 in the wavelength range of 4 μm to 20 μm [6]. Cao et al. [7, 8] designed ring-shaped and columnar metamaterial absorbers based on the characteristics of metal-dielectric-metal (MIM) multilayer structure, with the average absorptivity of 95% in the visible light

Y. Wei and S. Chng (Eds.): INCASE 2021, LNME, pp. 203–207, 2022.
https://doi.org/10.1007/978-981-16-5763-4_45

band. The broadband metamaterial absorber proposed has achieved an average absorption rate of over 90% with a peak value of 99.8% at the wavelength of 354 nm ~ 1066 nm [11]. An ultra-wideband high-performance metamaterial absorber was designed based on Fabry-Perot resonance effect and hybrid coupling theory, the absorption band of the absorber is 300 nm ~ 3000 nm, and the average absorptivity is more than 93% [12].

2 Simulation and Result Analysis

The simulation structure modeling is shown (see Fig. 1). Two simulation models: titanium convex structure (see Fig. 1(a)) and titanium convex structure covered with ZnO nanorods (see Fig. 1(b)). The finite-difference time-domain (FDTD) method (numerical solutions, Inc.) is used to solve the Maxwell's equations. The reflection spectrum of the absorber in the visible to near infrared (0.4 μm ~ 2 μm) region is calculated. The optical constants of titanium, ZnO and SiO_2 are obtained from Palik (Handbook of Optical Constants of Solids I-III by E. Palik). The incident light source is a plane light source with a wavelength of λ. The simulation structure is periodic and the simulation range is a single period. The X and Y axes are set as periodic boundary conditions and the Z axis is the PML ideal boundary condition. The simulation accuracy is λ/12. A suitable field power monitor is selected to monitor and analyze the reflection spectrum of the structure surface.

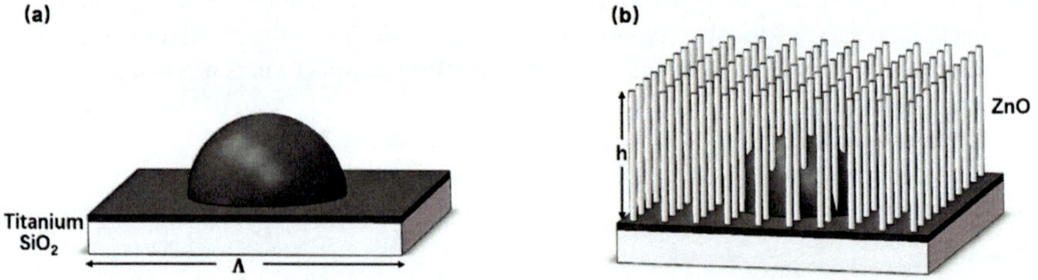

Fig. 1. Simulation Model. (a) Titanium convex structure. (b) Titanium convex structure covered with ZnO nanorods.

Based on the titanium convex array structure to study the influence of convex height (H) and period (Λ) on the reflectance spectrum (see Fig. 2(a) and Fig. 2(b)). The effect of the convex height on reflection spectrum at the diameter of 1μm and the period of 1.4 μm. It can be found that there is an obvious absorption peak at the wavelength of 1.4 μm (equal to the period of convex array structure). The higher the convex height is, the more obvious the absorption peak is, and there is a slight red shift. When the convex height is 0.4 μm, the absorptivity (1-reflectivity) at the wavelength of 1.4 μm can reach 98.6% (see Fig. 2(a)). The relationship between reflection spectrum and period at the diameter of 1 μm and the height of 0.5 μm. It can be found that with the increase of the period, the absorption peak has a slight red shift, and the corresponding wavelength of the absorption peak is close to the period. When the period is 1.9 μm, the maximum absorption is 100% at the wavelength of 1.9 μm (see Fig. 2(b)). Based on the titanium convex array structure covered with ZnO nanorods to show the effect of the height (H) and period (Λ) of titanium convex structure on the reflection spectrum with the diameter of ZnO nanorods is 0.1 μm, the height of ZnO nanorods is 1 μm and the spacing of ZnO nanorods is 0.05 μm (see Fig. 2(c) and Fig. 2(d)). It can be found that the total reflectivity decreases and the number of absorption peaks increases compared with the case of only titanium convex structure. The relationship between reflection spectrum and height, period is similar to that of pure titanium convex structure, but the regularity is not obvious.

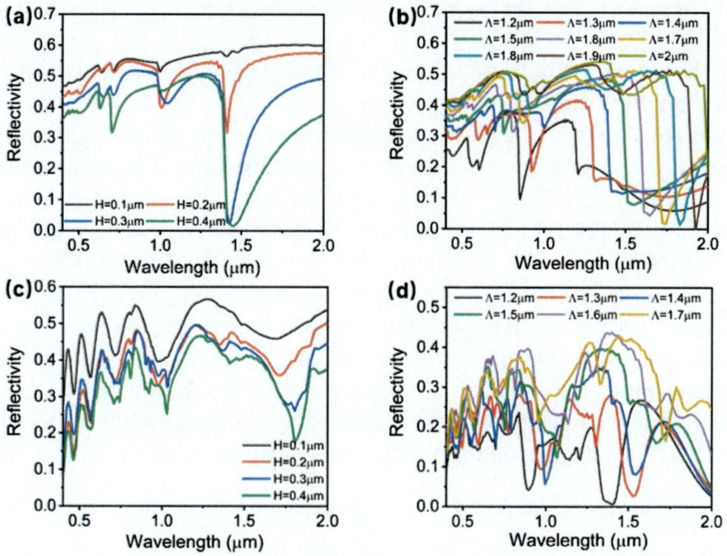

Fig. 2. The reflection spectrum of titanium convex structure. (a) Convex height (H). (b) Period (Λ). The reflection spectrum of titanium convex structure covered with ZnO nanorods. (c) Convex height (H). (d) Period (Λ).

Based on the titanium convex structure with ZnO nanorods to study the influence of height (h), diameter (d) and spacing (t) of ZnO nanorods on the reflection spectrum when the diameter of titanium convex structure is 1 μm, the height is 0.5 μm and the period is 1.4 μm (see Fig. 3). The effect of the diameter of ZnO nanorods on the reflection spectrum (see Fig. 3(a)). At the wavelength of 0.4 μm ~ 2 μm, the reflectivity decreases with the increase of diameter. There are two obvious absorption peaks (wavelength of 1μm and 1.4 μm). The absorption peak at the wavelength of 1.4 μm, with the increase of the diameter of ZnO nanorods, the absorption peak has a slight red shift. When the diameter is 0.1 μm, the corresponding maximum absorptivity of the two absorption peaks (wavelength of 1 μm and 1.52 μm) is about 95%. The effect of the height of ZnO nanorods on the reflectance spectrum (see Fig. 3(b)). It can be found that with the height of ZnO nanorods is larger, the overall reflectance is smaller, and at the wavelength of 1.6 μm the reflectance can reach 100%. The effect of the spacing of ZnO nanorods on the reflection spectrum(see Fig. 3(c)). The results show that the smaller the spacing is,

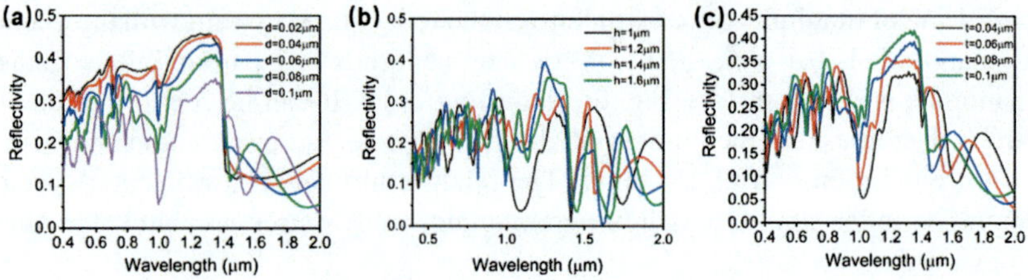

Fig. 3. Reflectance spectra of titanium convex array structure covered with ZnO nanorods. (a) The diameter of ZnO nanorod. (b) The height of ZnO nanorod. (c) The spacing of ZnO nanorod.

the smaller the reflectivity is, the more obvious the absorption peak is, and the absorption peak has a slight blue shift.

The changes of electric field and magnetic field in XZ plane based on the titanium convex array structure and the titanium convex array structure covered with ZnO nanorods are studied (see Fig. 4). The diameter of titanium convex is 1 μm, the height is 0.5 μm, the period is 1.4 μm, the diameter of ZnO nanorods is 0.1 μm, the height is 1 μm, and the spacing is 0.05 μm. The absorption peak of titanium convex array structure is due to the surface plasmon, which is a kind of electromagnetic excitation mode propagating at the interface of metal and medium. Because the wavelength of the absorption peak is equal to the period of the convex array structure, the changes of electric field and magnetic field in XZ plane of two structures at wavelength of 1.4 μm are studied. It can be found that the electric field changes obviously at the air and titanium convex structure, and the maximum electric field enhancement on the convex surface is 5 times.

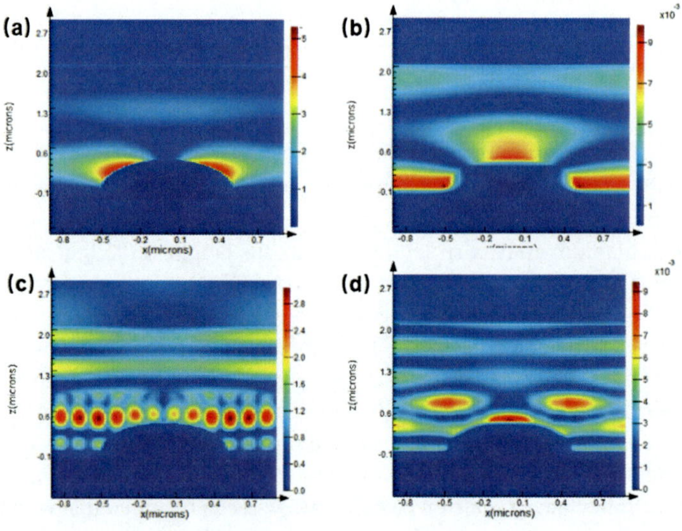

Fig. 4. The change of electric field and magnetic field in XZ plane at the wavelength of 1.4 μm. Titanium convex array structure. (a) Electric field. (b) Magnetic field. Titanium convex array structure covered with ZnO nanorods. (c) Electric field. (d) Magnetic field.

3 Conclusion

The influence of different structural parameters on the reflectance spectra of titanium convex array structure is obtained numerically. It is concluded that the wavelength of the absorption peak is equal to the period of the convex array structure, which is attributed to the surface plasmon effect. For the specific structure parameters, the absorption can reach 100% at the specific wavelength. In addition, the change of reflection spectrum is compared when the convex array structure is covered with ZnO nanorods. It is found that the absorptivity and the number of absorption peaks increase. The influence of structural parameters of ZnO nanorods on the reflectance spectra is also studied.

Acknowledgement. We acknowledge the financial support from the National Natural Science Foundation of China (NSFC) with Grant Nos. 51875006 and 51705009, and the Key Cultivation Projects of BJUT.

References

1. Hao, J., Wang, J., Liu, X., Padilla, W., Zhou, L., Qiu, M.: High performance optical absorber based on a plasmonic metamaterial. Appl. Phys. Lett. **96**(25), 251104 (2010). https://doi.org/10.1063/1.3442904
2. Shen, X.: Polarization-independent wide-angle triple-band metamaterial absorber. Opt. Express **19**(10), 9401–9407 (2011)
3. Shen, X.P.: Dual band metamaterial absorber in microwave regime. Acta Physica Sinica **61**(5), 58101 (2012)
4. Hedayati, M.K.: Design of a perfect black absorber at visible frequencies using plasmonic metamaterials. Adv. Mater. **23**(45), 5410–5414 (2011)
5. Cui, Y.: Ultra-broadband light absorption by a sawtooth anisotropic metamaterial slab. Nano Lett. **12**(3), 1443–1447 (2011)
6. Wang, H.: Perfect selective metamaterial solar absorbers. Opt. Express **21**(106), 1078–1093 (2013)
7. Cao, S.: Broadband efficient light absorbing in the visible regime by a metananoring array. Ann. Phys. **526**(1–2), 112–117 (2014)
8. Cao, S.: Two-dimensional subwavelength meta-nanopillar array for efficient visible light absorption. Appl. Phys. Lett. **102**(16), 1611009 (2013)
9. Ramya, S.: A compact ultra-thin ultra-wideband microwave metamaterial absorber. Microw. Opt. Technol. Lett. **59**(8), 1837–1845 (2017)
10. Lu, Z.: Design and absorption characteristics of broadband nano-metamaterial solar absorber. Acta Optica Sinica **37**(9), 0923001 (2017)
11. Lei, L.: Ultra-broadband absorber from visible to near-infrared using plasmonic metamaterial. Optics. Expr. **26**(5), 5686 (2018)
12. Yang, W.A.: Numerical study of an ultra-broadband, wide-angle, polarization-insensitive absorber in visible and infrared region. Optical. Mater. **114**(3), 110902 (2021)

A Simple Waviness Evolution Model for the Stream Finishing Process

Stephen Wan[1](✉), Shengwei Ma[1], Cary Turangan[1], Keni Chih-Hua Wu[1], Sho Itoh[2], Wei Shin Cheng[3], and Kai Liang Tan[3]

[1] Institute of High Performance Computing, 1 Fusionopolis Way, #16–16, Connexis, Singapore, Singapore
{wansym,masw,cary,keni-wu}@ihpc.a-star.edu.sg
[2] Graduate School of Engineering, Faculty of Engineering, Chiba University, 1-33 Yayoi-cho, Inage-ku, Chiba-shi, Chiba 263-8522, Japan
ittosyo@chiba-u.jp
[3] Advanced Remanufacturing and Technology Centre, 3 Cleantech Loop, #01/01, CleanTech 2, Singapore, Singapore
{chengws,tan_kai_liang}@artc.a-star.edu.sg

Abstract. We present the development of a waviness evolution model for the stream finishing process. The model, derived along the same lines as that of a previously developed surface roughness evolution model, was tested against waviness experimental data gathered by processing coupons additively coated with stainless steel 410 by laser metal deposition. To complete the study, continuum-based granular flow dynamics simulations were carried out to estimate the media pressure loading on and velocities across the coupons in order to uncover the trends observed in the experimental data. While the model was initially targeted at the stream finishing process, it appears that it is also applicable to other media finishing processes, as there was no loss of generality in the derivation of the model.

Keywords: Stream finishing · Waviness model · Stainless steel coupons · Granular flow dynamics CFD simulations

1 Introduction

The present work is a preliminary attempt at developing a simple model for the purpose of understanding the evolution of the surface waviness of parts undergoing the stream finishing process.

Stream finishing is a relatively new mass finishing process. It may be considered to be a variant of the better-known drag finishing in that, instead of forcing workpieces through a stationary volume of media, workpieces are processed by immersing them in a rotating drum of media.

There appears to be a dearth of work, published in the open literature, related to stream finishing and none that includes surface waviness. For example, Schulze et al. [1] studied the effect of processing time on the micro surface hardness, residual stresses,

Y. Wei and S. Chng (Eds.): INCASE 2021, LNME, pp. 208–211, 2022.
https://doi.org/10.1007/978-981-16-5763-4_46

and roughness of AISI 4140 steel specimens. This study was later extended by Karacas et al. [2] to include to the effect of drum rotational speed. Zanger et al. [3] undertook a combined experimental and numerical analysis for a deeper understanding of the relationships between various process parameters and normal forces, material removal, and surface topography for AISI 4140 specimens.

2 Methods

2.1 Waviness Evolution Model

We assume that the waviness profile is sinusoidal in shape and, hence, consists of protuberances generated by revolving a sinusoidal cross-section, which can be simplified to a triangular one, described by an equation of the form (Fig. 1):

$$h_p = c - mr \tag{1}$$

in which, c may be interpreted as the height of the asperity and m defining its width. Then, developing along the same lines as the derivation of the roughness, R_a evolution model, outlined in Wan et al. [4], we arrive at the waviness, W_a, evolution model:

$$W_a = W_\infty - \frac{\sqrt[3]{\frac{3m^2 k_W p_g v_s}{H} t + [a_W(W_\infty - W_i) + c]^3} - c}{a_W} \tag{2}$$

where W_i is the initial waviness; W_∞, the limiting waviness; p_g, the pressure loading; v_s, the sliding velocity; H, the hardness of the material being worn; and a_W, a geometrical scaling factor.

2.2 Experimental Set-up

To test the waviness model, waviness data was obtained by processing a coupon coated with stainless steel 401 by direct laser deposition, immersed in a stream fishing drum to media depth, $z = 150\text{mm}$ and radial arm length, $r = 250\text{mm}$; and oriented at a glancing angle, $\theta_z = 45^\circ$ and zero tilt angles, θ_1 and θ_2. The drum was rotated at speed, $R_1 = 30$ RPM (Fig. 2). Triangular, prismatic shaped ceramic DBS 4X4 media (sized $d = 4\text{mm}$, density$\rho_p = 2083\text{kgm}^{-3}$) were used in the experiments.

2.3 Simulation Set-up

Computational fluid dynamics (CFD) simulations based on a continuum-based granular flow dynamics model were carried out to estimate the media pressure, p_g loading on and velocities, v_s sliding over the test coupons. The simulation setup is outlined in Wan et al. [5] and will not be repeated here.

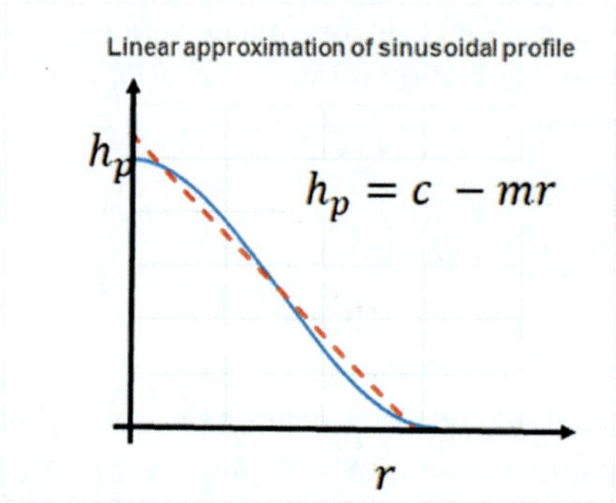

Fig. 1. Linear approximation of protuberance solid of revolution cross-section.

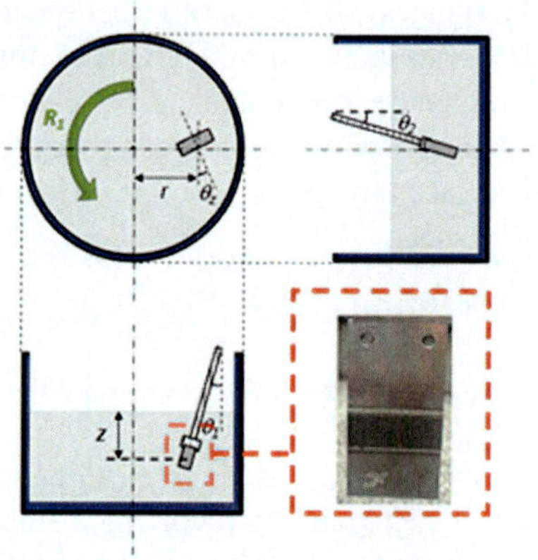

Fig. 2. Experimental setup Inset: test coupon.

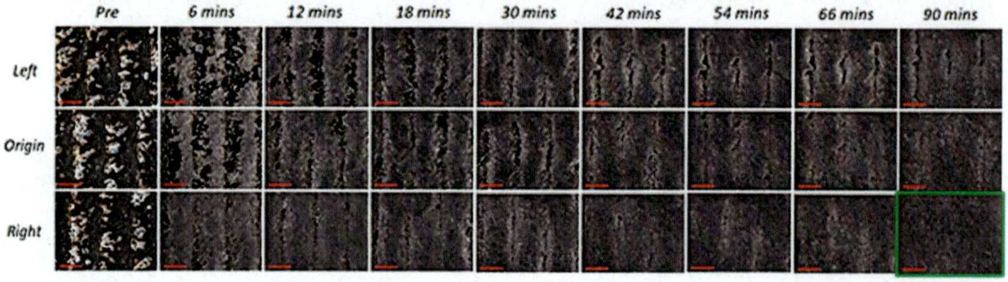

Fig. 3. Pictographs of left, centre (origin) and right sections of the test piece at various processing times

3 Results and Discussion

The pictograph of the test piece surface at various processing times (Fig. 3) indicates, visually, that material removal occurred at the greater rate from the left end towards the right end of the test piece. This is reflected in the waviness results, as shown in Fig. 4.

Interestingly, the polishing strength = $p_g v_s$, as estimated from the CFD simulations, also increases from left to right, as shown in Fig. 5. This would help explain the trends in the experimental and model-predicted data, assuming the various constants in the model equations carry the same values, looking at the model equation (Eq. 2).

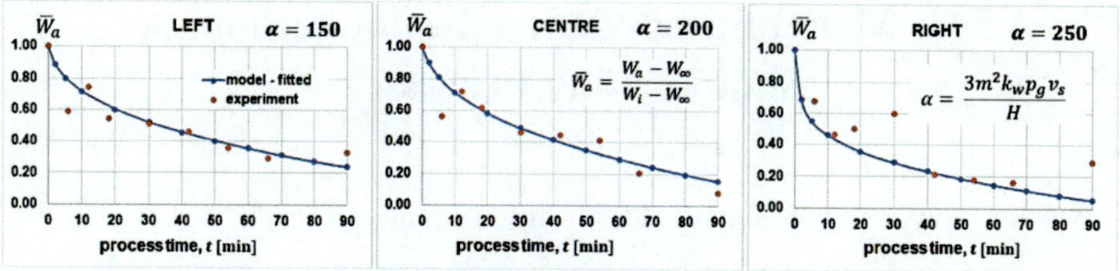

Fig. 4. Experimental waviness data and model-fitted curves for $\alpha = 150, 200, 250$.

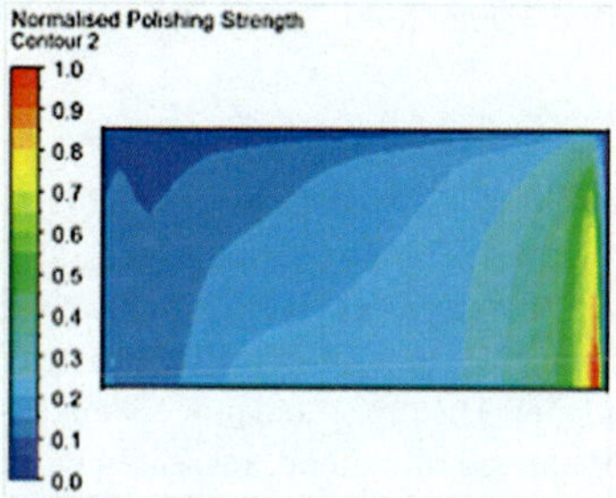

Fig. 5. Polishing strength distribution over test piece surface.

4 Conclusions

The waviness evolution model (Eq. 2), together with the CFD simulation estimation of the polishing strength, appears to be able to physically explain the trends in the experimental waviness results. Furthermore, while the model was initially targeted at the stream finishing process, it appears that it is also applicable to other media finishing processes, as there was no loss of generality in the derivation of the model.

References

1. Schulze, V., Gibmeier, J., Kacaras, A.: Qualification of the stream finishing process for surface modification. CIRP Ann. Manuf. Tech. **66**, 523–526 (2017)
2. Kacaras, A., Gibmeier, J., Zanger, F., Schulze, V.: Influence of rotational speed on surface states after stream finishing. Procedia CIRP **71**, 221–226 (2018)
3. Zanger, F., Kacaras, A., Neuenfeldt, P., Schulze, V.: Optimization of the stream finishing process for mechanical surface treatment by numerical and experimental process analysis. CIRP Ann. Manuf. Tech. **68**, 373–376 (2019)
4. Wan, S., Liu, Y., Woon, K.S., Tnay, G.L.: A material removal and surface roughness evolution model for loose abrasive polishing of free form surfaces. Int. J. Abrasive Tech. **6**(4), 269 (2014)
5. Wan, S., Ma, S., Turangan, C., Wu, C.K., Itoh, S., Cheng, W.S., Tan, K.L., Ho ,J.: Calibration of a continuum-based granular flow dynamics model for the numerical simulation of the stream finishing process, INCASE 2021. (submitted)

A PID Model Based Controller for Automated Shot Peening Processes

Van Bo Nguyen[1(✉)], Te Ba[1], A. Teo[2], Ampara Aramcharoen[2], K. Ahluwalia[2], and Changwei Kang[1]

[1] Institute of High Performance Computing, A*STAR, Singapore S138632, Singapore
nguyenvb@ihpc.a-star.edu.sg

[2] Advanced Remanufacturing and Technology Centre, A*STAR, Singapore 637143, Singapore

Abstract. This study reports a newly developed model-based controller for automated shot peening processes. The proportional-integral-derivative (PID) controller is developed based on process model to provide an optimal control action for process control, while the reference set-point is translated from the target setting intensity using a proxy model. The proxy and process models are developed using machine learning and optimization method, respectively. The developed control system is demonstrated using both in-silico and onsite controls. The results show that the control performance is accurate, stable, and robust. It implies that the developed control system can be applied to practical applications for reduction of cost, time, and material wastage.

Keywords: PID control · Shot peening · Peening intensity

1 Introduction

Shot peening technology is widely used in the surface enhancement applications to improve the fatigue life of the components by introducing a thin layer of compressive residual stress near surface through multiple impacts [1]. In shot peening, the peening intensity is an important output parameter in addition to the coverage. The intensity represents the kinetic energy of the shot stream transferred to workpiece [2]. The intensity and corresponding exposure time are obtained from the saturation curve, which is often plotted using at least four measurement points of different peening times [3]. In industrial production, development of the Almen system often takes about 20–25% of production effort to establish process parameters for a desired intensity. Thus, an advanced control system for shot peening process is needed to upgrade the existing control system for fully automated operation and minimising the development effort. Thus, it can greatly reduce the total cost, time, and labour for future peening operations.

The application of PID control technology has shown a great reduction in the cost, time, and labour of many manufacturing productions [4, 5]. As such, a practical PID control technology is a suitable choice to upgrade the current control technology shot peening process to a smarter operation. In PID controller designing and tuning, there are two direct and indirect common methods often employed [6]. For the direct method,

Y. Wei and S. Chng (Eds.): INCASE 2021, LNME, pp. 212–216, 2022.
https://doi.org/10.1007/978-981-16-5763-4_47

controller parameters are directly fine-tuned using feedback data during operation. This requires a determination of a suitable characteristics of the relevant control system properties, and proper ways of modifying the controller parameters to obtain the objective(s) [6]. Therefore, the direct method often requires a long period of time to fine-tune the process model and controller. In the indirect method, the control parameters are obtained from a process model that is developed based on data model, analytical, or theoretical model [7]. The indirect (or model based) control performance is often robust, accurate and stable as the model-based control can be applied for the auto-tuning of control parameters. Thus, in this study, we choose the model-based control technology to develop the control system for a smarter shot peening machine.

2 Methodology

2.1 Process Model and Proxy Model

In this study, a single input-single output (SISO) control system is developed based on the process model and proxy model. The proxy model is developed to translate the target intensity into the pressure reference set-point at the nozzle for real-time tracking and feedback control. The process model is developed to link the inlet air pressure to the pressure at nozzle for controller development. To develop the proxy model, a total of 40 operational conditions (4 values of inlet air pressure (20, 30, 44 and 54 psi), and 5 values of media flowrate (1, 2, 3, 4 and 5 kg/min)) experimental trials are performed for both the ASR 70 and ASR 230 peen types to collect the measurement data of intensity and air pressure. Each operational condition is repeated at least 4 different exposure times to obtain saturation curve, intensity, and exposure time. The XGBOOST machine learning algorithm [8] is employed to develop proxy model by training 80% of the measurement data, while rest 20% of measurement data is used for validation (test). The results of proxy model are shown in Fig. 1 (left).

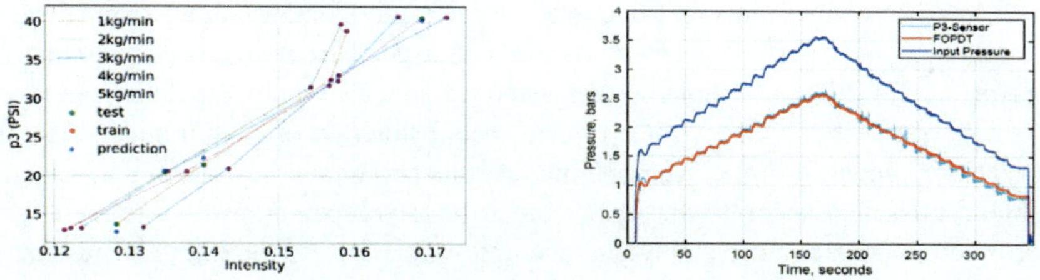

Fig. 1. Results of proxy model (left) and process model (right).

To build the process model, a total of 10 experimental trials using 5 values of media flowrate (1, 2, 3, 4 and 5 kg/min) are performed for increment and decrement in inlet air pressure from 20 to 54 psi with a step of 2 psi for both peen types to collect the air pressure in time series. The optimization method is applied to the measurement data to obtain the model parameters in the form of first order plus dead-time model with control ($\tau_p \frac{dy(t)}{dt} = -y(t) + K_p u(t - \theta_p)$) [9]. Figure 1 (right) shows the comparison results of the model and measurement data for the case of 3 kg/min and ASR 230 media type. It implies that the obtained model is accurate and can exactly represent the dynamics of the shot peening process.

2.2 Control Framework Development

The SISO control system is developed and implemented based on the internal process model for auto-tuning of the control parameter with the change of the process dynamics. A python OPC-UA package is employed to transfer the data between control system and physical shot peening machine. The PID controller is often expressed in the form of $u(t) = K_p e(t) + K_I \int_0^t e(\tau) d\tau + K_d \frac{de(t)}{dt}$. The control parameters (K_p, K_I and K_d) are evaluated using model parameters (K_p, τ_p, and θ_p) with K_C, $K_I = K_p/\tau_i$, and $K_d = K_C \tau_d$ [6, 9]. The developed control system is shown in Fig. 2 (left), while the integration diagram is shown on the Fig. 2(right).

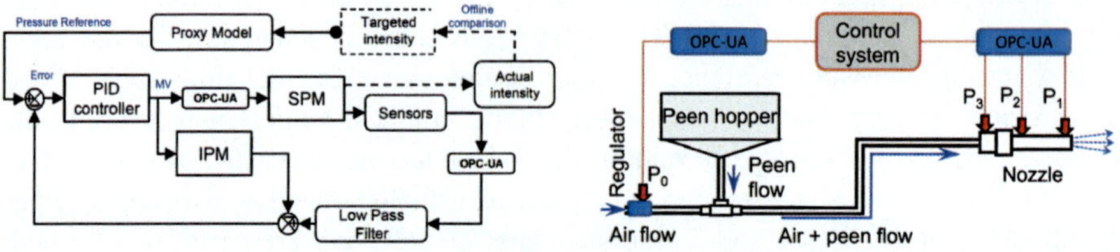

Fig. 2. The control system architecture for shot peening machine (left) and the schematics of the integration of the control system into the shot peening machine (right).

3 Results and Discussion

Both the in-silico control and on-site control are performed for the validation and demonstration of the control performance and accuracy. A total of 10 in-silico control scenarios are performed for different media flowrates and reference set-points of both peen types. The results of in-silico control for a typical media flowrate of 3 kg/min of ASR 230 peen type are shown in Fig. 3. The obtained results show that the control performance is stable, robust, and accurate. In other words, the developed control system can give an accurate control action to stably and accurately manipulate the inlet air pressure to achieve the different pressure reference set-points.

For the on-site control scenarios, both control performance and accuracy are tested for both peen types. A total of 30 test cases are carried out at the physical shot peening machine. In which, the low, medium, and high intensities are tested for each media flowrate. Like the in-silico control, the obtained results of on-site control also show that the developed control system can accurately execute the control actions to drive the dynamical process to the setting value. The maximum error of the measurement and setting intensity is about 0.006mmA, which is smaller than the industrial threshold of ± 0.01 mmA [3]. Figure 4 shows a typical on-site control results of the case with media

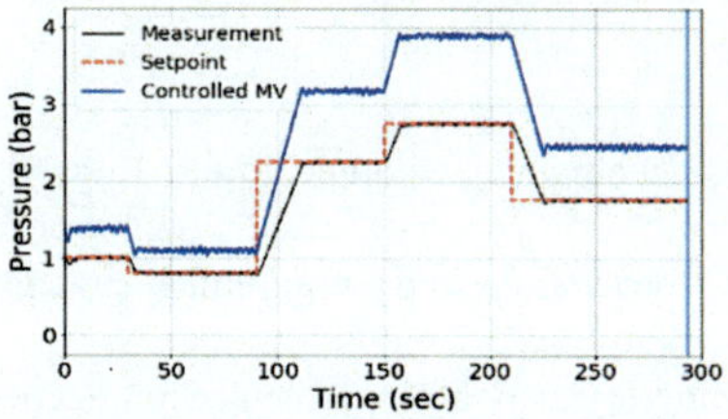

Fig. 3. Results of in-silico control performance for different pressure set-points of the ASR 230 peen type and media flowrate of 3 kg/min.

flowrate of 4 kg/min for ASR 70 peen type. The real-time tracking pressure is shown in Fig. 4 (left), while the measurement intensity and exposure time is shown on the right. The target setting intensity is 0.139 mmA, while the obtained measurement is 0.138 mmA.

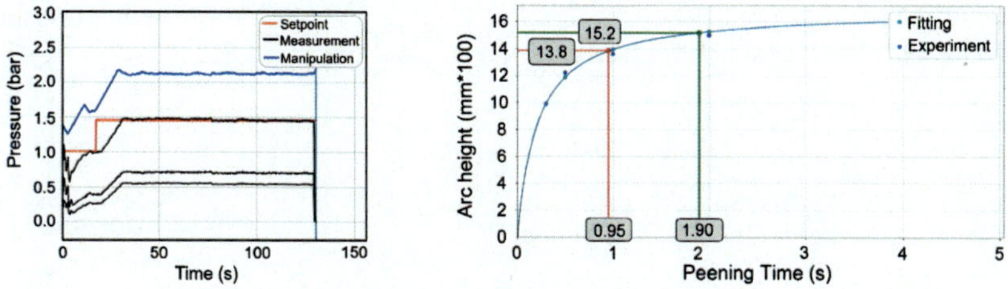

Fig. 4. Real-time pressure reference tracking (left) and measurement intensity (right).

4 Conclusions

In this study, the model-based control system is developed for a smarter and fully automated shot peening process. This developed control system includes an internal process model, a PID controller, auto tuning control parameters, a noise filtering technique, and a proxy model. The control system is integrated into a physical shot peening machine for on-site control demonstrations. Performance of the developed control system is demonstrated using both the in-silico and on-site control for different scenarios. Both in-silico and on-site control demonstrations show that the control performance is stable, robust, and accurate. In another word, this internal process model-PID control system can significantly reduce the cost, time, and labor, as well as material waste as compared to the development using Almen system for actual operations.

Acknowledgements. This project is supported by the A*STAR Science and Engineering Research Council (SERC), under its RIE2020 Advanced Manufacturing and Engineering IAF-PP Program (Grant No: A1894a0032).

References

1. Marsh, K.J.: Shot peening: techniques and applications. Eng. Mater. Adv. Ser. Ltd. United Kingdom (1993)
2. Kirk, D.: Peening intensity true meaning and measurement strategy. Shot Peener **30**(3), 26–30 (2016)
3. Champaigne, J.: Shot Peening Intensity Measurement. Shot Peener **6**(4), 1–6 (1992)
4. Xia, C., et al.: A review on wire arc additive manufacturing: monitoring, control and framework of automated system. J. Manuf. Sys. **57**, 31–45 (2020)
5. McMillan, G.K.: Industrial Applications of PID Control in Advances in Industrial Control. Springer, London (2012)
6. Åström, K.J.: Theory and applications of adaptive control-a survey. Auto. **19**(5), 471–486 (1983)
7. Rivera, D.E., Morari, M., Skogestad, S.: Internal model control: PID controller design. Ind. Eng. Chem. Proc. Des. Dev. **25**(1), 252–265 (1986)
8. Chen, T., Carlos, G.: XGBoost: A Scalable Tree Boosting System. In: International Conf. on Knowledge Discovery and Data Mining, ACM Press (2016)
9. Brunton, S., Kutz, J.: Data-Driven Science and Engineering: Machine Learning, Dynamical Systems, and Control. Cambridge University Press, Cambridge (2019)

Phase-Field Crystal Modelling of Deformation Behavior in Grain Growth

Yi Jiang[1,2], Xiling Yao[2], and Seung Ki Moon[1(✉)]

[1] School of Mechanical and Aerospace Engineering, Nanyang Technological University, 50 Nanyang Ave, Singapore 639798, Singapore
yjiang025@e.ntu.edu.sg, skmoon@ntu.edu.sg

[2] Singapore Institute of Manufacturing Technology, 73 Nanyang Dr, Singapore 637662, Singapore
yao_xiling@simtech.a-star.edu.sg

Abstract. Simulation methods have been developed to study material microstructure and manufacturing process. In phase-field modelling, a phase-field crystal method is a powerful tool for microstructural evolution simulation across multiple scales. Using the parameters fitted for a specific material, the actual material property can be simulated mathematically. In this study, the deformation behavior during grain growth is simulated using a phase-field crystal simulation method. Body-centred cubic Fe is selected as a case study. Simulation results show that strain affects the form and length of grain boundaries and the structure of the grains, which can be explained in the energy perspective as extra energy obtained by the system from the strain. The study indicates the potential application of the phase-field crystal simulation method in the additive manufacturing industry, which will reduce cost and time and help design new material.

Keywords: Additive manufacturing · BCC Fe · Deformation behavior · Phase-field crystal

1 Introduction

The microstructure evolution process of a metal material is of importance to their physical properties and mechanical properties. As the improvement of conventional phase-field simulation, a phase-field crystal (PFC) simulation method introduced by Elder et al.[1], can be applied to a multiscale simulation algorithm that bridges the classical molecular dynamics (MD) simulations and the conventional phase-field simulation methods. The PFC method introduces an order parameter that can produce the "periodicity" of crystal lattices [2]. During rapid solidification processes such as additive manufacturing (AM), there will be residual stress. The highly localized heat source, its motion, and the relatively slow heat conduction induce thermal gradients in AM materials, resulting in significant residual stresses [3]. The objective of this study is to analyse how deformation affects the grain growth of metal, with an example of body-centred cubic (BCC) structure Fe.

Y. Wei and S. Chng (Eds.): INCASE 2021, LNME, pp. 217–221, 2022.
https://doi.org/10.1007/978-981-16-5763-4_48

2 PFC Simulation Process

The PFC model has been developed using Ginzburg-Landau theory in the phenomenological approach. Hence, the evolution of locally conserved order parameter φ is given by

$$\frac{\partial\varphi(r)}{\partial t} = M\nabla^2\frac{\delta F}{\delta\varphi(r)} \tag{1}$$

where r is the position vector, M is the mobility, and F is the free energy functional. The one-mode free energy functional chosen here is stated as

$$F = \int_V \left\{\frac{1}{2}\varphi(r)\left[\alpha + \lambda\left(q_0^2 + \nabla^2\right)^2\right]\varphi(r) + \frac{g}{4}\varphi(r)^4\right\}dr \tag{2}$$

where α, λ, q0, and g are phenomenological parameters that fit the specific properties of the specific material. Extra experiments, simulations, and calculations can be conducted to obtain the parameters. For computational purposes, the dimensionless form of the free energy can be obtained by substituting the following terms:

$$\varepsilon = -\frac{\alpha}{\lambda q_0^4}, \psi = \varphi\sqrt{\frac{g}{\lambda q_0^4}}, x = q_0 r, F' = \frac{g}{\lambda^2 q_0^5}F, t' = Mtq_0^3. \tag{3}$$

Then the equation turns to:

$$F' = \int_V \left\{\frac{1}{2}\psi\left[-\varepsilon + \left(1 + \nabla^2\right)^2\right]\psi + \frac{1}{4}\psi^4\right\}dx \tag{4}$$

In this equation, F' represents dimensionless free energy functional, and ε is a constant representing temperature, as it is proportional to the deviation of the temperature from the melting point. The parameter ψ represents the dimensionless atom density field.

The parameters of BCC Fe is obtained from Asadi et al. [4]. The phenomenological parameters for a specific material can be determined by MD using the modified embedded-atom method (MEAN), are calculated as,

$$\alpha = \frac{2575k_BT}{132}, \lambda = \frac{10k_BT}{5.47224}, g = 66.2786k_BT, q_0 = 3.02.$$

For the convenience of simulation, φ is set to be -0.0036, so that the result is set in the stable triangular phase region. The simulation is performed by the semi-implicit Fourier spectral algorithm using MATLAB®.

Aiming to study the influence of the axial deformation during AM process, a strain of 0.2% was applied in the vertical direction. To simulate the deformation in a two-dimensional coordinate system, the altered grid size V of the model was used with chosen strain rate $\dot{\varepsilon}$. While keeping the constant volume, we have

$$V = dxdy = dx'dy' = (dx + \dot{\varepsilon}\Delta t)dy' \tag{5}$$

3 Simulation Result and Discussion

The simulation was run in the area of a 512 × 512 grid as shown in Fig. 1. Two series of simulations were conducted for the different initial condition of nuclei. The directions of nucleuses 1, 3, and 5 were the same, and the directions of nucleuses 2, 4, and 6 were the same. For Series 1, the direction of grain 1 is perpendicular to the direction of grain 2, and for Series 2, the angle is 75°. The central positions of each nucleuses were 1(128, 384), 2(128, 128), 3(448, 448), 4(448, 344), 5(448, 240), 6(448, 136) with the origin set in the bottom left corner.

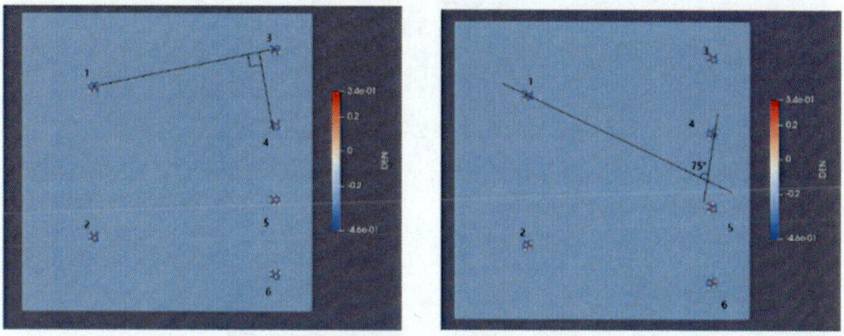

Fig. 1. The initial grain position.

During grain growth, the area of each grain will increase with time. In the position that only influenced by one grain, the direction of the new generated region will not change, and in the position that influenced by more than one grain, there will be an interaction, which both grains affect the new generated region and try to make it the same direction of itself. Finally, the interaction reached an equilibrium state and formed grain boundaries (see Fig. 2). Grain boundaries between different directions were marked and measured with green color.

Firstly, grain boundary length under strain is more extended than without strain for two cases in this study. As the axial strain applying to the material, the energy amount is larger than the case without applied strain. For Series 1, the length is 1387.9 units without strain and 1407.5 units with strain. For Series 2, the length is 1487.1 units without strain and 1561.2 units with strain. One possible explanation could be about energy. As axial strain applying to the material, the energy put into the material is larger than the case without applied strain. More energy leads to a more extensive grain boundary length, as grain boundary is the area with high free energy. Secondly, the grain boundary interface between different directions formed differently with and without strain. This can be explained by the cause of deformation, which leads to different forms at the interface region. Thirdly, according to the density field variables analysis as shown in Fig. 3, an evident dislocation can be seen since there is a gap in the top area of simulation Series 1 without strain, which does not happen with strain, indicating the applied strain helped eliminate grain boundary in the area with the same direction. However, it may generate new defects, as shown in the bottom part of Fig. 2B.

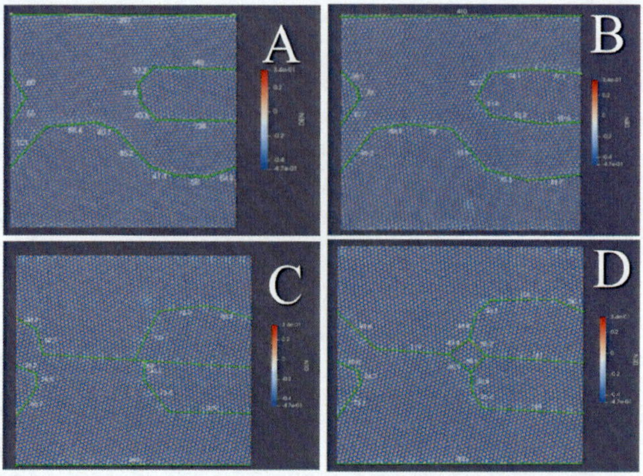

Fig. 2. Final simulation results. A: Series 1 without strain; B: Series 1 with strain; C: Series 2 without strain; D: Series 2 with strain.

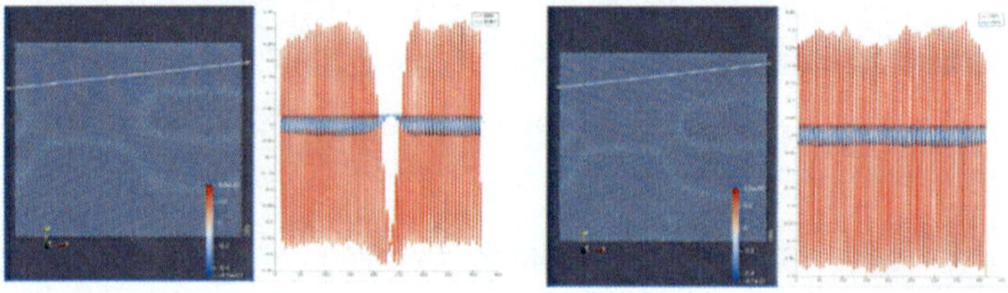

Fig. 3. The density field variables analysis for simulation Series 1. The left diagram is without strain, and the right diagram is with strain.

4 Conclusions

In this research, PFC simulation was conducted to analyze the microstructure behavior with parameters that fit BCC iron material. This study focused on the deformation during AM, indicating axial strain on microstructure and explaining it from an energy perspective. PFC simulation in AM process can contribute to a timesaving and resource-saving data-driven material and process design.

Acknowledgements. This research is supported by a grant from Korea Institute of Material Science and Singapore Centre for 3D Printing.

References

1. Elder, K.R., Grant, M.: Modeling elastic and plastic deformations in nonequilibrium processing using phase field crystals. Phys. Rev. E **70**(5), 051605 (2004)
2. Biner, S.B.: Phase-field crystal modeling of material behavior. In: Biner, S.B. (ed.) Programming Phase-Field Modeling, pp. 337–368. Springer, Cham (2017). https://doi.org/10.1007/978-3-319-41196-5_7

3. Denlinger, E.R., Heigel, J.C., Michaleris, P., Palmer, T.A.: Effect of inter-layer dwell time on distortion and residual stress in additive manufacturing of titanium and nickel alloys. J. Mater. Process. Technol. **215**, 123–131 (2015)
4. Asadi, E., Asle Zaeem, M., Baskes, M.I.: Phase-field crystal model for Fe connected to MEAM molecular dynamics simulations. JOM **66**(3), 429–436 (2013). https://doi.org/10.1007/s11837-013-0845-3

Check for updates

Design Consideration for Additively Manufactured Components Through Topology Optimization and Generative Design for Weight Reduction

Jian-Yuan Lee[1,2], Arun Prasanth Nagalingam[1,2], and Swee Hock Yeo[1,2](✉)

[1] School of Mechanical and Aerospace Engineering, Nanyang Technological University, 50 Nanyang Avenue, Singapore 639798, Singapore
MSHYEO@ntu.edu.sg
[2] Rolls-Royce@NTU Corporate Lab, N3.1-B2a-01, 50 Nanyang Avenue, Singapore 639798, Singapore

Abstract. Additive manufacturing is an emerging advanced manufacturing technology and provides a new digital manufacturing tool with higher degree of design freedom for designer. However, there are some potential technical challenges, such as lack of part screening methodology as well as design consideration for additively manufactured components. Therefore, the main objective of this paper is to propose a part screening methodology via a two-steps approach for reducing of the weight of test chamber components used as a case study. Two designs for additive manufacturing methodologies: (1) topology optimization; (2) generative design was used to optimize components and to achieve up to 50% weight reduction. Some further design consideration and recommendations are also provided for the future applications for various industries.

Keywords: Additive manufacturing · Generative design · Part screening · Part consolidation · Surface finishing

1 Introduction

Additive manufacturing (AM) is an advanced digital manufacturing technology with many novel material options [1] and potential applications in various industries, such as in aerospace [2], and water technologies [3]. There are several advantages of using AM for serial production, especially during the COVID-19 period [4], such as smarter supply chain with on-demands and cost-efficient production of spare parts. In addition, the designers can have higher degree of freedom of AM components via topology optimization or generative design. On the other hand, test chamber is a critical component of the hydrodynamic cavitation abrasive finishing (HCAF) process [5–9], that aids to achieve the target surface textures (i.e. Ra < 1 μm and Rz < 20 μm) in additively manufactured components. However, there are some technical challenges, such as the poor surface quality for as-built AM component [10]. Therefore, the main objectives of this

Y. Wei and S. Chng (Eds.): INCASE 2021, LNME, pp. 222–225, 2022.
https://doi.org/10.1007/978-981-16-5763-4_49

paper are to reduce of the weight of selected test chamber components and develop a part screening methodology via a two-step approach. The main objective of these optimization methods is to remove unnecessary material for weight reduction while maintaining the structural integrity and functional requirements. Last but not least, some important design consideration and surface finishing for additively manufactured components will be briefly highlighted at the last section as well.

2 Methods

2.1 Part Screening Methodology

The part screening methodology is mainly via a two-step approach: (1) Step 1: High-level feasibility study and (2) Step 2: detailed technical evaluation.

Step 1: High-Level Feasibility. The main purpose of high-level level feasibility is to classify the traditional manufactured parts into three categories: (1) machined parts; (2) welded parts; and (3) standard parts. There are currently 120 test chamber components in total, such as the standard parts (e.g. O-ring), machined parts (e.g. top flange, manifold) and welded parts (chamber housing and door hinge). There are about 63% machined parts, 20% of welded parts, and 17% of standard parts based on our part classification of the current test chamber components.

Step 2: Detailed Technical Evaluation. After completion of high-level feasibility study, the four high potential HCAF test chamber components were further evaluated for part consolidation and design for AM. The main purpose of part consolidation is to identify the maximum number of components that be combined together. Next, two common design for AM strategies, generative design and topology optimized were used to optimize the current design for weight reduction without compromising the mechanical properties of the HCAF test chamber component as shown in Fig. 1.

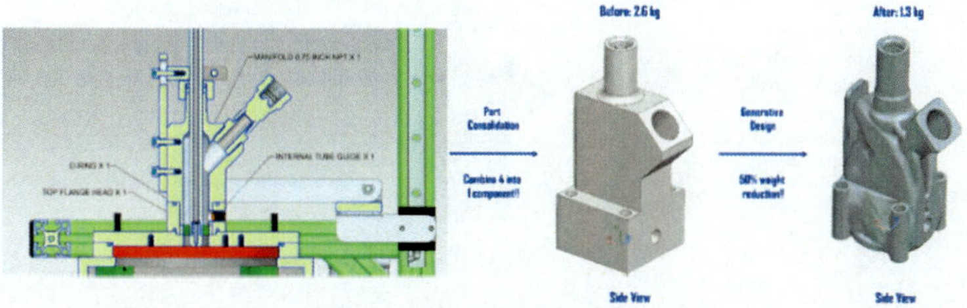

Fig. 1. Part consolidation and topology optimization via generative design

3 Results and Discussion

3.1 Part Consolidation and Generative Design

The total number of selected test chamber components was reduced from 4 components into 1 component via part consolidation. For generative design, the key design input parameters are the offset thickness of shell and the applied load on the manifold for the design space. The offset thickness is 2.5 mm with the constraint to ground and it was fixed at the bottom during topology optimization and generative design. The optimization methodology is to minimize strain energy subject to mass target and the weight target was fixed at 1.3 kg as one of the optimization constraints. The total weight reduction is about 50% from 2.6 kg to 1.3 kg after several iterations of topology optimization via generative design. Even though the total weight can be further reduced to as low as 0.6 kg but there is a trade-off between the weight and mechanical properties due to the thinner wall thickness and more material removal. In addition, aluminum alloy was selected as the final AM material due to the high strength-to-weight ratio and lower cost (S$ 1,270) and shorter printing time (10 h) at the same layer thickness.

3.2 Design Consideration and Surface Finishing

After several iterations of topology optimization and generative design, it is critical to consider the printability and design considerations as well. There are a few important design considerations, such as build orientation, support removal, design for additive manufacturing, and post-processing to print the optimized component via laser-based powder bed fusion. For example, vertical build orientation was selected to minimize the effort on support removal. In addition, all the horizon holes were designed for additive manufacturing from circular geometry to either diamond or teardrop geometry to prevent the distortion in a large lateral hole with diameter larger than 4 mm as shown in Fig. 2.

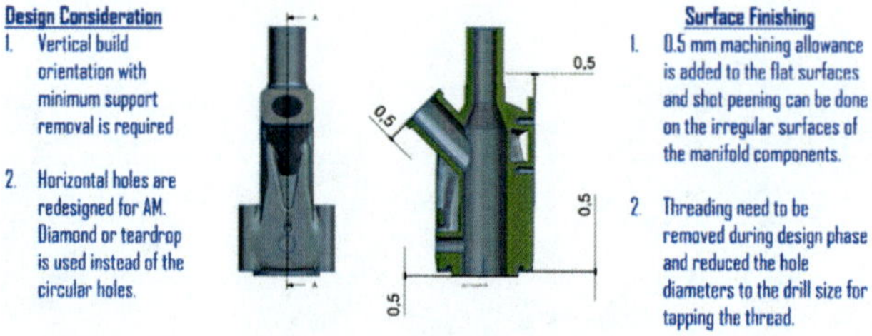

Fig. 2. Design consideration and surface finishing for additively manufactured test chamber components.

For external surface finishing, a 0.5 mm machine allowance is added to all the flat surfaces as well as the bottom surfaces for machining while shot peening can be done on the irregular surfaces of the optimized manifold components.

4 Conclusion

In summary, a case study of design consideration and surface finishing for additively manufactured HCAF test chamber components via topology optimization and generative design for weight reduction was presented in this paper. A part screening methodology two-steps approach was used to select the 4 high-potential components for part consolidation out of the total 120 test chamber components. Two designs for additive manufacturing method: topology optimization and generative design were employed to reduce the weight of the additively manufactured components and up to 50% of the component weight can be reduced without affecting the performance, structural integrity and mechanical properties of the test chamber.

References

1. Lee, J.-Y., An, J., Chua, C.K.: Fundamentals and applications of 3D printing for novel materials. Appl. Mater. Today **7**, 120–133 (2017)
2. Uriondo, A., Esperon-Miguez, M., Perinpanayagam, S.: The present and future of additive manufacturing in the aerospace sector: a review of important aspects. Proc. Inst. Mech. Eng. Part G: J. Aerosp. Eng. **229**(11), 2132–2147 (2015)
3. Lee, J.-Y., et al.: The potential to enhance membrane module design with 3D printing technology. J. Membr. Sci. **499**, 480–490 (2016)
4. Choong, Y.Y.C., et al.: The global rise of 3D printing during the COVID-19 pandemic. Nat. Rev. Mater. **5**(9), 637–639 (2020)
5. Nagalingam, A.P., Yuvaraj, H.K., Yeo, S.H.: Synergistic effects in hydrodynamic cavitation abrasive finishing for internal surface-Finish enhancement of additive-manufactured components. Addit. Manuf. 101110 (2020)
6. Nagalingam, A.P., Yuvaraj, H.K., Santhanam, V., Yeo, S.H.: Multiphase hydrodynamic flow finishing for surface integrity enhancement of additive manufactured internal channels. J. Mater. Process. Technol. 116692 (2020)
7. Nagalingam, A.P., Yeo, S.H.: Surface finishing of additively manufactured Inconel 625 complex internal channels: a case study using a multi-jet hydrodynamic approach. Addit. Manuf. **36**, 101428 (2020)
8. Nagalingam, A.P., Yeo, S.H.: Controlled hydrodynamic cavitation erosion with abrasive particles for internal surface modification of additive manufactured components. Wear **414–415**, 89–100 (2018)
9. Nagalingam, A.P., Thiruchelvam, V.C., Yeo, S.H.: A novel hydrodynamic cavitation abrasive technique for internal surface finishing. J. Manuf. Process. **46**, 44–58 (2019)
10. Lee, J.-Y., Nagalingam, A.P., Yeo, S.H.: A review on the state-of-the-art of surface finishing processes and related ISO/ASTM standards for metal additive manufactured components. Virtual Phys. Prototyp. 1–29 (2020)

Simulation Study on High Order Harmonic Generation by Ultra-intense Laser Driven Grating Targets

Yang Wang[1,2](✉), Haiying Song[1,2], Bing Wang[1,2], Xun Liu[3], Wei Li[3], and Shibing Liu[1,2]

[1] Strong-Field and Ultrafast Photonics Lab, Faculty of Materials and Manufacturing, Beijing University of Technology, Beijing 100124, China
ynztwy_mst@emails.bjut.edu.cn, {hysong,sbliu}@bjut.edu.cn

[2] Key Laboratory of Trans-Scale Laser Manufacturing Technology, Ministry of Education, Beijing 100124, China

[3] Beijing Institute of Space Mechanics and Electricity, Beijing 100094, China

Abstract. We theoretically study high-order harmonic generation (HHG) from relativistically driven overdense plasma targets with rectangularly grating-structured surfaces by femtosecond laser pulses. Particle-in-cell (PIC) simulations show that, intense harmonic emissions along target surface are produced. These harmonics linked to the coherent radiations from each grating protuberance. The intense near-surface harmonics is observed due to the strong dependence of harmonic components on the incident angle. The spatial distribution of harmonics is mainly dominated by the oscillation of equivalent electric-dipole (OEED) and spatial modulation. The intensities of harmonic emission are stronger than that in the case of a normal incidence, which is associated with the orientation of equivalent electric dipole.

Keywords: Ultra-intense laser-matter interaction · High-order harmonic generation · Surface grating structure · Surface enhancement · Particle-in-cell simulation

1 Introduction

High-order harmonic generation (HHG) by intense ultrashort laser pulses impinging solid targets is of great potential to produce bright coherent radiations. So far it has been extensively studied both theoretically and experimentally [1]. Improving the efficiency of harmonic emission from solid-density plasma targets is one of the essential purposes [2]. In recent years, the target shaped grating surface has been proposed to enhance the harmonic intensity [3].

In this report, by using Particle-In-Cell simulation, we investigate the HHG from a laminar grating target irradiated by an oblique incident ultrashort laser pulse. The near-surface harmonic emissions are coherently enhanced within a relatively wide range of the incident angles. Especially, the near-surface harmonic emissions at oblique incidence

Y. Wei and S. Chng (Eds.): INCASE 2021, LNME, pp. 226–229, 2022.
https://doi.org/10.1007/978-981-16-5763-4_50

are much more intense than those at normal incidence, which can be attributed to the orientation of oscillation of Equivalent electric-dipole (OEED) formed by the electron bunches and immobile targetions.

2 Methods

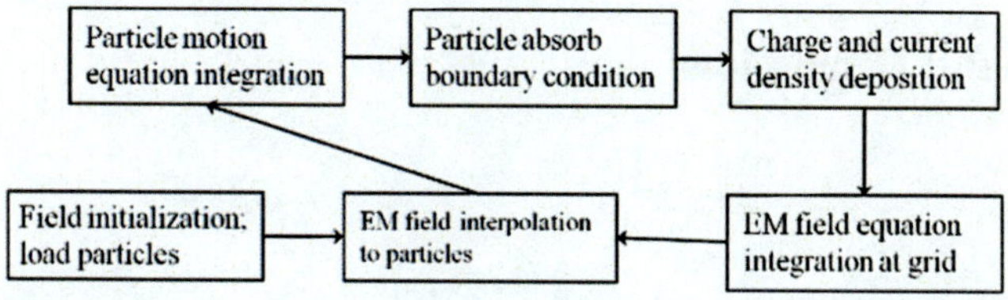

Fig. 1. Flow diagram of Particle-in-Cell simulation scheme.

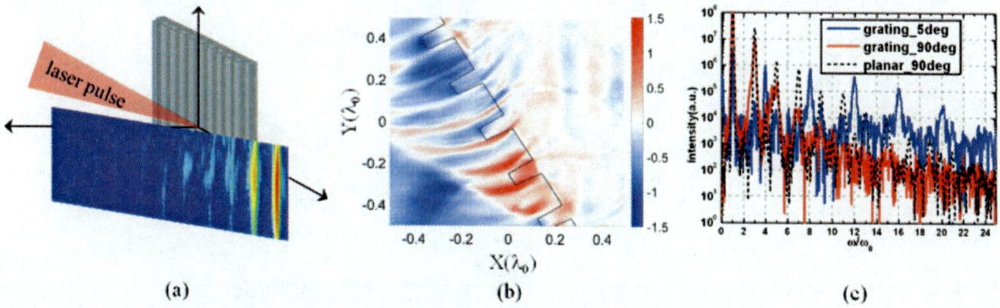

Fig. 2. (a) Laser-grating target interaction scheme, here the incident angle is chosen as 30o, (b) electric field normal to the target surface distribution during interaction, (c) harmonic spectra of grating target and planar target (for comparison).

The Particle-In-Cell simulation scheme is presented in Fig. 1. Electromagnetic (EM) field is solved by Maxwell equations:

$$\nabla \times \mathbf{E} = -\frac{\partial B}{\partial t} \tag{1}$$

$$\nabla \times \mathbf{B} = \mu_0 \mathbf{J} + \mu_0 \epsilon_0 \frac{\partial E}{\partial t} \tag{2}$$

These EM field equations are numerically solved by standard FDTD algorithm. And the motion of each plasma particle is solved by the equation:

$$\frac{d\mathbf{p}_i}{dt} = q_i(\mathbf{E}(t, \mathbf{r}_i) + \mathbf{v}_i(t) \times \mathbf{B}(t, \mathbf{r}_i)) \tag{3}$$

It is numerically solved by the three-step scheme [4]. The contribution of each particle to current density is deposited to the discrete spatial grids, which is used to update the EM field.

Simulations are based on the software VSim [5]. Grating material is replaced by overdense plasma due to the rapid ionization by ultrashort laser pulse. Periodicity $d = (1/4)\lambda_0$, protuberance width $e = (1/16)\lambda_0$ and height $h = (1/8)\lambda_0$. Laser wavelength λ_0 is chosen at 800 nm, polarization is perpendicular to grooves. The transverse and temporal profiles are all Gaussian forms with beam width and pulse duration at 1.6 μm and 10.4 fs. Laser intensity is set as 5×10^{18}W/cm^2. Plasma density is set as $N_e = 20N_c$, where $N_c = \varepsilon_0 m_e w_0^2/e^2$ $N_c = \epsilon_0 m_e \omega_L^2/e^2$ is the critical plasma density.

3 Results and Discussions

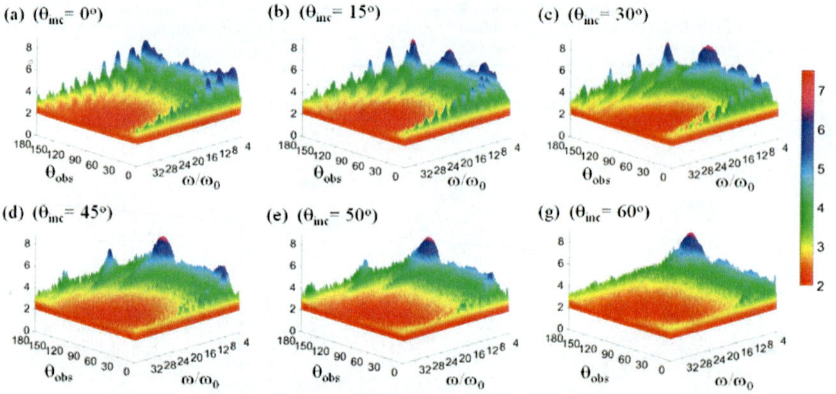

Fig. 3. The angular distribution of emission intensity with harmonic order at the incident angles of 0°, 15°, 30°, 45°, 50° and 60°. (θ_{inc}: incident angle; θ_{obs}: observation angle; ω/ω_0: radiation field frequency normalized by laser frequency ω_0)

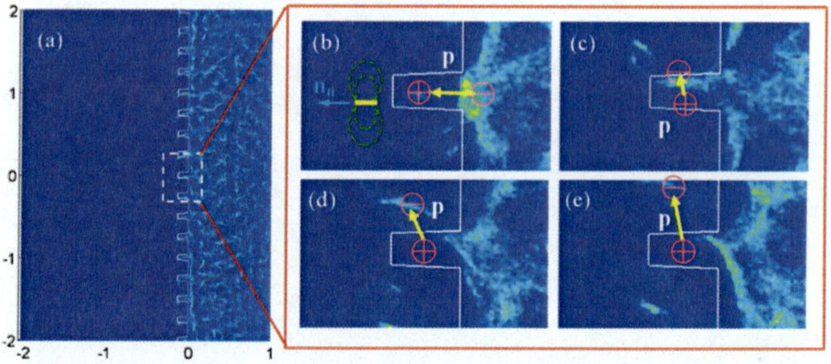

Fig. 4. Surface electron density evolution of laser pulse interacting with target at different time. (a) Electron density of whole target, (b–e) Electron density of central protuberance.

As shown in Fig. 2(c), for grating target, harmonics along the small angle (5°) with their harmonic orders at 4, 8, 12, 16 and 20 obviously stand out. Figure 2(b) shows the electric field spatial distribution during the interaction. An obvious high-frequency field component between reflection wavefront and target surface is propagating along the target surface direction.

Figure 3 shows the angular distributions of harmonic emission. With the increase of incident angle the intense near-surface harmonics appear. For large incident angle as 60°, Harmonic energy almost confined with a small angle region near 150° observation angle, which means that these harmonics can be collected in experiment. Angular distribution of harmonic emission shows up significant difference in large and small angles of incidence condition.

Surface electrons (surface plasmon) suffer from electric field force and ponderomotive force by laser pulse, while the target ions are treated as an immobile positively charged background in ultra fast interactions. The transient equivalent electric dipole is formed between the electronegative center and electropositive center. The strong electrostatic force between the centers attempts to regain its initial electroneutrality, leading to the oscillation of the equivalent electric-dipole and harmonic emission by relativistical dipole radiation with energy flux density [6]:

$$\mathbf{S} = \frac{|dp/dt|^2 \sin^2\theta}{32\epsilon_0\pi^2 c^3 R^2}\mathbf{n}_0 \tag{4}$$

Each electron bunch-protuberance pair corresponds to one dipole oscillator and the radiation field of this oscillator at radius R conforms to "8-shaped" spatial distribution as illustrated in Fig. 4. The "dipole radiations" are modulated by the periodical grating structure thus the harmonic intensities are significantly enhanced.

4 Conclusion

Harmonic emissions from grating targets rely on the collective motion of surface electrons, which results in the formation of equivalent electric-dipoles and their collective oscillations. Harmonic radiant orders are selectivity with different incident angles which is relative to the orientation of the oscillating dipole. For large incident angles, harmonic energy is well confined within a small angle region.

Acknowledgements. This work is supported by the National Key Research and Development Program (No. 2018YFB0504400), the Beijing Education Committee Science and Technology Plan (KM201810005001).

References

1. Teubner, U., Gibbon, P.: High-order harmonics from laser-irradiated plasma surfaces. Rev. Mod. Phys. **81**, 445–479 (2009)
2. Tsakiris, G.D., Eidmann, K., Meyer-ter-Vehn, J., Krausz, F.: Route to intense single attosecond pulses. New J. Phys. **8**, 19 (2006)
3. Fedeli, L., Sgattoni, A., Cantono, G., Macchi, A.: Relativistic surface Plasmon enhanced harmonic generation from gratings. Appl. Phys. Lett. **110**, 051103 (2017)
4. Birdsall, C.K., Langdon, A.B.: Plasma Physics via Computer Simulation. McGraw-Hill Book Company, New York(1985)
5. Nieter, C., Cary, J.R.: Vorpal: a versatile plasma simulation code. J. Comput. Phys. **193**, 448 (2004)
6. Jackson, J.D.: Classical Electrodynamics, 3rd edn. Wiley, Hoboken (1999)

Development of Shot Peening Coverage Evaluation Model from 3D Topography Analysis on Ti-6Ai-4V

Yu Han Aldrich Chua(✉) and Dawei Xu

Advanced Remanufacturing and Technology Centre, 3 Cleantech Loop, #01/01, CleanTech 2, Singapore, Singapore
{aldrich-chua,xudw}@artc.a-star.edu.sg

Abstract. Coverage is a key quality parameter in the shot peening process. The current method to quantify coverage is highly subjective as it relies on operators' visual judgment and experience. Claims have been made to use three-dimensional (3D) topography to access coverage, but the detailed methodology is still under investigation. In this study, Abbott Firestone curve was used to access the 3D topography of Ti-6Ai-4V samples; and a correlation model was established to assess coverage. High correlation was obtained from validations and optimizations, which reduced uncertainties from the conventional manual inspections.

Keywords: Shot peening · Coverage · 3D topography · Abbott Firestone curve

1 Introduction

Coverage is a critical parameter in the shot peening process control, and it is described as the percentage of a peened area over the component's total surface area. The current quantification method is subjective due to the reliance on operators' visual judgement.

A few studies have raised the idea of using surface topography to correlate with coverage. One method suggested the use of 2D surface measurements from a profilometer, to access the profile differences between low and high coverage specimen [1]. However, high variance was shown through a study [2] as a 2D topography is from a linear profile rather than an area. Another method suggested 3D surface topography from interferometry or confocal microscopy: Thompson et al. proposed the study of impact dent characterization between peened and unpeened surfaces to assess the coverage [3]. The method is similar in terms of filtering the height profile from the unpeened region from a total peened area to determine the coverage percentage. However, no sufficient experimental results or analysis were indicated in the proposed methods, and hence it was unclear how indentations created by shot peening were differentiated.

In surface topography, the Abbott-Firestone curve is a plot that shows the cumulative distribution of profile heights over a measured area. This provides a better understanding of distribution of surface peaks and valleys [4], and could potentially differentiate

Y. Wei and S. Chng (Eds.): INCASE 2021, LNME, pp. 230–234, 2022.
https://doi.org/10.1007/978-981-16-5763-4_51

between a peened and unpeened area of a surface. Therefore, this study looks into finding a link between the Abbott-Firestone curve to coverage, and in turn providing a solution to measure coverage percentage on a shot peened surface.

2 Experimental Methods

Reference specimens (1 batch with 5 pieces/batch) were used to first build correlation models between the Abbott-Firestone curve and coverage. After which, the models were validated using the validation specimens peened in this study. Validation specimens (5 batches) were peened with different intensities, media types and time (where T_{100} is the time required to achieve 100% coverage) to study if the generated models were adaptive enough to detect coverage for different parameters. Predicted coverage from the models were then compared against a coverage assessment conducted by 6 shot peening experts on the validation specimens. This was to done to assess the prediction accuracy of the models against the current method of quantifying coverage. Peening parameters are as shown in Table 1.

Table 1. Shot peening parameters applied to specimens

Specimens type	Batch number	Media type	Intensity (mmA)	Coverage
Reference specimens	1	ASR230	0.23	40%, 60%, 70%, 80%, 90%, 95%, 100%, 125%
Validation specimens	1	ASR230	0.30	$0.10\ T_{100}$, $0.30\ T_{100}$, $0.40\ T_{100}$, $0.50\ T_{100}$, $0.75\ T_{100}$, $1.00\ T_{100}$, $1.25\ T_{100}$, $2.00\ T_{100}$
	2	ASR230	0.30	
	3	ASR230	0.15	
	4	ASR70	0.25	
	5	ASR70	0.40	

The Abbott-Firestone curve was generated from 2 mm by 2 mm 3D topography profiles at different locations of the same sample using a confocal microscope (Fig. 1(a)). As shown in Fig. 1(b), for a 60% reference specimen's Abbott-Firestone curve, the coverage (y value: 60% of area whose position is deeper than the remaining regions) and the corresponding normalized cumulative height (x value) were correlated as a point-of-interest (POI). After access all the reference specimens, the corresponding POIs were used to fit a correlation equation (Fig. 1(c)): each specimen was scanned on 3 locations and hence 3 corresponding POIs were obtained; the correlation function was fit from their mean coverage values. This equation was then used to validate the coverage from the validation specimens.

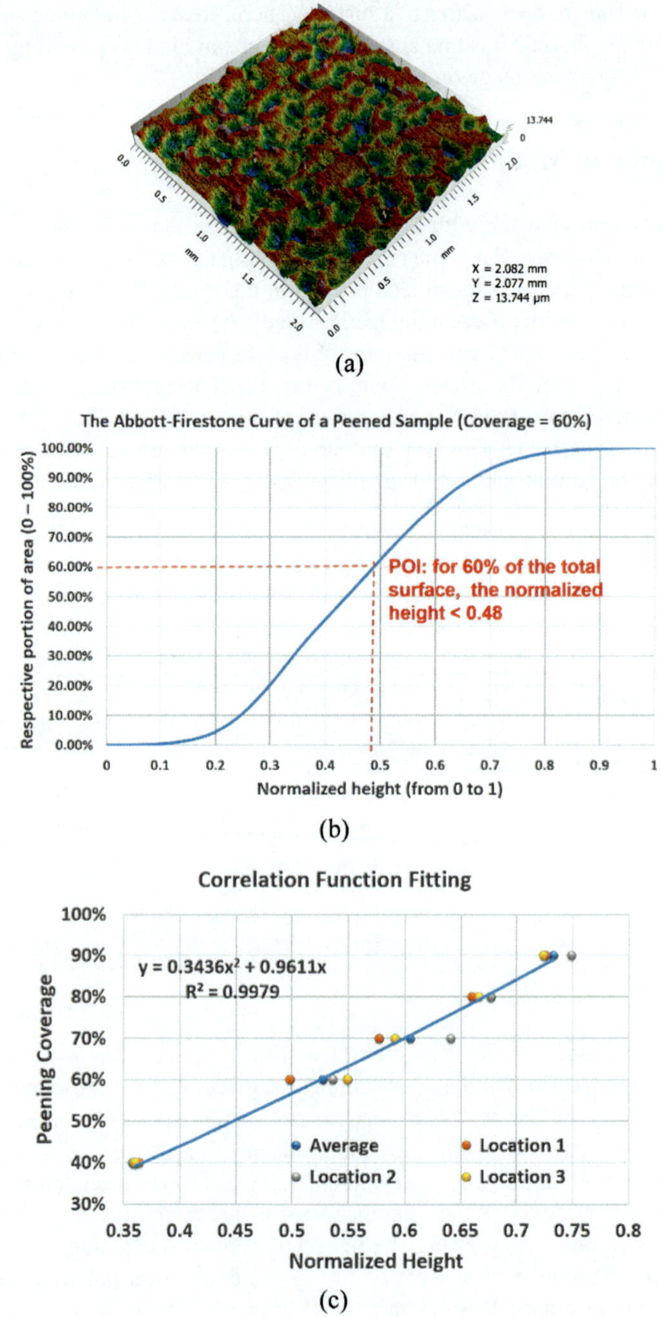

Fig. 1. 3D topography obtained (a) with the Abbott-Firestone curve at 60% coverage (b), compiled data points from Abbott-Firestone curve to create a correlation curve (c).

After the validation specimens were peened, each specimen was sent for 3D topography measurements to generate their respective Abbott-Firestone curves. Each curve was then plotted together with the correlation equation, as the example shown in Fig. 2: a specimen was test via the correlation function, and the output coverage is ~40%. The intersection (between the specimen's Abbott-Firestone curve and the correlation function) was obtained, and its Y-axis value would be the calculated coverage percentage for the test specimen (~40%).

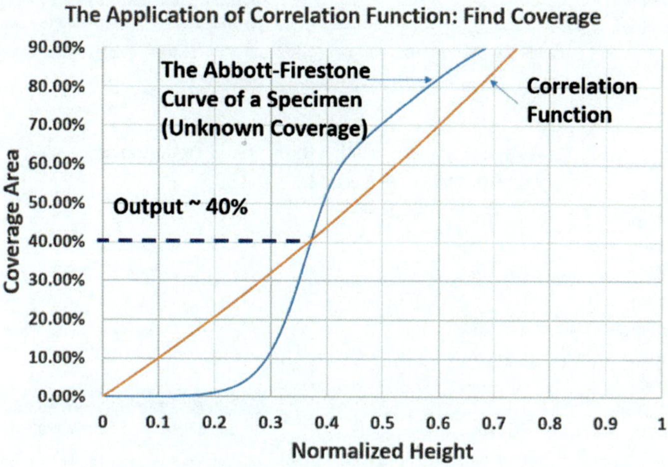

Fig. 2. Using the correlation equation to find coverage from the Abbott-Firestone curve

3 Experimental Results and Discussions

From the first validation (40 samples (5 batches and 8 pieces per batch)), ±10% deviation (model prediction vs. the reference (experimental expert assessment)) was obtained from only 23 of them. To further improve the prediction accuracy, additional 3 batches of specimen (24 pieces) with difference peening conditions, was added as reference training data to further optimize the correlation function. As shown in Fig. 3, the updated model was able to predict 35 out of 40 total measurements with the average deviation between the predicted coverage and the measurements of 2.08% (Fig. 3). There were some instances (5 out of 40 pieces) where the equation created was unable to predict the coverage values within ±10% from the experts' reference. One suspected reason is that both reference and validation specimen's coverage readings are based on operators' visual inspection, which is the conventional subjective approach with uncertainties. Another possible factor is related to the area of assessment; the test area (3D topography) was only a 2 mm by 2 mm region, whereas the experts uses the entire specimen, which may also introduces inspection deviations.

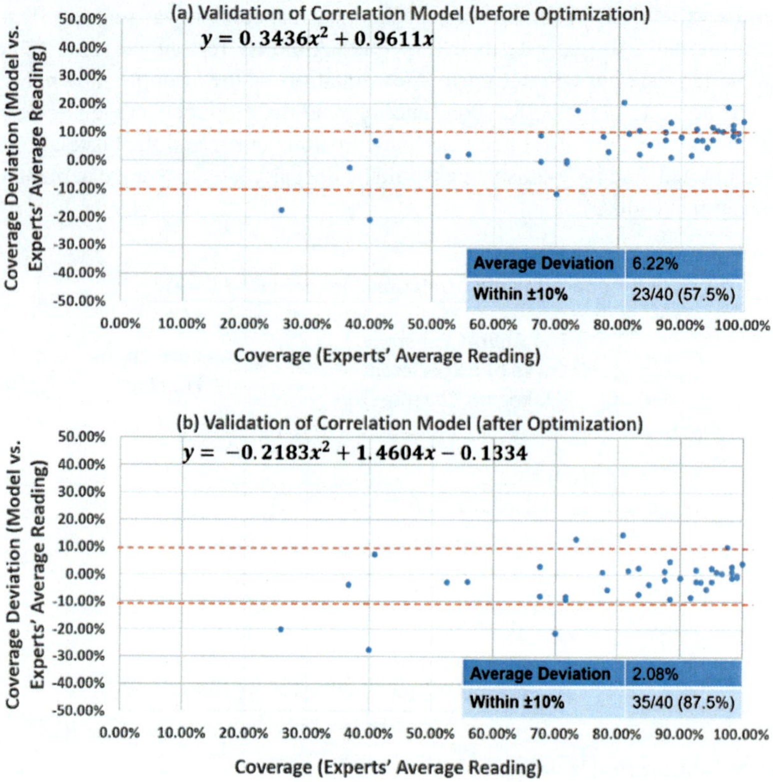

Fig. 3. Model vs. manual inspection: (a) before and (b) after optimization.

4 Conclusions

This study used Abbott Firestone curve to create a correlation model between 3D topography and shot peening coverage. The model was able to predict 35/40 measurements with an average deviation of 2.08% from reference. The involved uncertainty attributes to the subjective inspection in both reference and validation specimen, and the area difference between 3D topography and the visual inspection.

References

1. Thompson, R.A., Tascillo, M.A.: U.S. Patent No. 5,235,517, U.S. Patent and Trademark Office, Washington, DC (1993)
2. Thompson, R.A., Tascillo, M.A.: U.S. Patent No. 5,293,320, U.S. Patent and Trademark Office, Washington, DC (1994)
3. Feldmann, G.G., Hennig, W.: U.S. Patent Application No. 13/009,188 (2011)
4. Coba-Salcedo, M., Coral, I., Valencia, G.: Characterization of surface topography with Abbott Firestone curve. Contemp. Eng. Sci. **11**, 3397–3407 (2018). https://doi.org/10.12988/ces.2018.87319

Effect of Coefficients of Restitution on the Particle Dynamics in Cold Spray and Shot Peening

Te Ba[1](✉), Changwei Kang[1], Ampara Aramcharoen[2], Zheng Zhang[3], Wiwik Karlina[4], Kannan Sundaravadivelu[1], Stephen Yee Ming Wan[1], Van Bo Nguyen[1], and Jisheng Pan[3]

[1] Institute of High-Performance Computing, A*STAR, Singapore 138632, Singapore
bat@ihpc.a-star.edu.sg
[2] Advanced Remanufacturing and Technology Center, A*STAR, Singapore 637143, Singapore
[3] Institute of Materials Research and Engineering, A*STAR, Singapore 138634, Singapore
[4] Institute of Infocomm Research, A*STAR, Singapore 138632, Singapore

Abstract. Effects of coefficients of restitution (COR) on the particle fate are evaluated numerically using Ansys Fluent. Two applications namely shot peening and cold spray are selected and particle velocities and spatial distribution at a certain distance are computed. Gas flow is computed based on the Navier-Stokes equation and then particles are injected into the gas flow to calculate the interaction between the gas flow and the particles. CORs are varied for both normal and tangential components to create a parametric study matrix. The variation of COR plays important role in the shot peening process but a minor role in the cold spray process from the analysis of the particle dynamics in the simulation.

Keywords: Coefficients of restitution · Shot peening · Cold spray · CFD simulation

1 Introduction

Both shot peening and cold spray technologies utilize high pressurized gas and powders (or shots) to enhance surface but in different industrial applications. Shot peening process induces a thin layer of compressive residual stress at the surface of the component via multiple peen impacts with high velocity. As for cold spray, the powders impact on the target component surface at a much higher velocity (over 500 m/s) than the shot peening process (around 100 m/s). At such a high impact velocity, instead of rebounding from the component surface, the powders deform and adhere to the surface to form a dense coating with excellent properties. Therefore, they share some common physics although they differ in the way to enhance the component surface.

In both processes, propellant gas is pressurized through a nozzle to create high velocity gas plume, where particles are injected and accelerated by the gas until they reach the target surface. In each equipment which may consist of components such as

Y. Wei and S. Chng (Eds.): INCASE 2021, LNME, pp. 235–239, 2022.
https://doi.org/10.1007/978-981-16-5763-4_52

chambers, pipes, hose, and nozzle, the particles may have possibilities to impact on the passage inner wall and rebound to get back to the streamed flow. In physical world, the impact/rebound is neither perfectly elastic nor perfectly plastic, which results in energy loss [1]. The coefficient of restitution (COR) is a method to benchmark such rebound behaviour and it is usually defined as the ratio of the object relative velocity right after and before the rebound [2, 3]. In practical and also adopted by Ansys Fluent [4], the value can be expressed in terms of its two components, tangential and normal coefficients. In this article, different combinations of normal and tangential coefficients are proposed to investigate the effect of them on the particle dynamics.

2 Methods

2.1 Experimental Settings and Measurement

The facility for cold spray is Plasma Giken's PCS-1000 system with a de Laval nozzle with axial injection and a throat in dimension of Φ3 mm. The nozzle used for the shot peening is straight nozzle with exit diameter 14 mm. The operating conditions for the cold spray are nitrogen gas at operating pressure 6 MPa and temperature 200 °C. For shot peening, the gas is compressed air at pressure 1.75 bar.

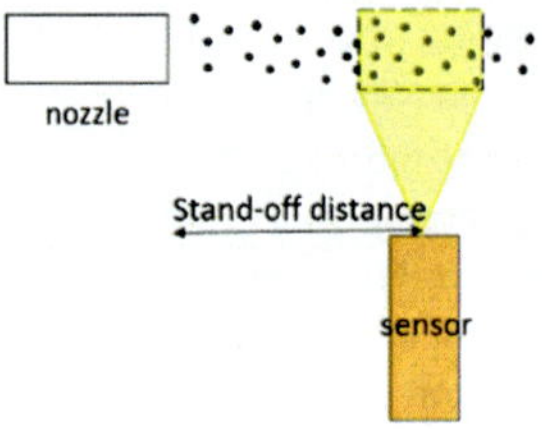

Fig. 1. Measurement of in-flight powders velocity using sensors.

Different from cold spray, where the injector is only ~70 mm upstream of the nozzle, there is a hose of ~15 m in-between the injector and the nozzle for shot peening facility. Optical sensors are used to measure the velocity of in-flight powders, namely cold spray meter for cold spray (at stand-off distance 30 mm) and shot meter for shot peening (at stand-off distance 150 mm), as illustrated in Fig. 1.

2.2 Modelling and Data Evaluation

Computational fluid dynamics (CFD) models are developed, whose computational domain includes components as the chamber, injector, nozzle, and a large air domain around the nozzle. In addition, hose is included for shot peening as it is the main passage of the peens. A group of constant normal and tangential COR (coefficients of restitution) is proposed as the DPM (Discrete Phase Model) boundary properties, which are summarized in Table 1. In addition, a pair of impact angle dependent CORs are also introduced for shot peening process, which is more realistic. The polynomial formula in terms of impact angle, as shown in Eq. (1) and (2), are fitted from a series of rebound modelling using LSDYNA. More details of the LSDYNA model are not depicted as that is out of current focus.

$$e_t = 1 - 0.721099\theta + 0.575020\theta^2 - 0.529610\theta^3 + 0.191170\theta^4 \quad (1)$$

$$e_n = 1 - 2.186530\theta + 2.732650\theta^2 - 1.679350\theta^3 + 0.402560\theta^4 \quad (2)$$

3 Results and Discussion

3.1 Effect of COR on Particle In-Flight Behavior in Cold Spray

The probability histogram and spatial distribution of the particles, captured at the stand-off distance of 30 mm, are shown in Fig. 2. The mean velocity of the in-flight particles computed from simulation is 587.6 m/s, which agrees well with the measured value 584.1 m/s. It is found that no matter which values the COR components are, the particle velocity and spatial distribution does not change.

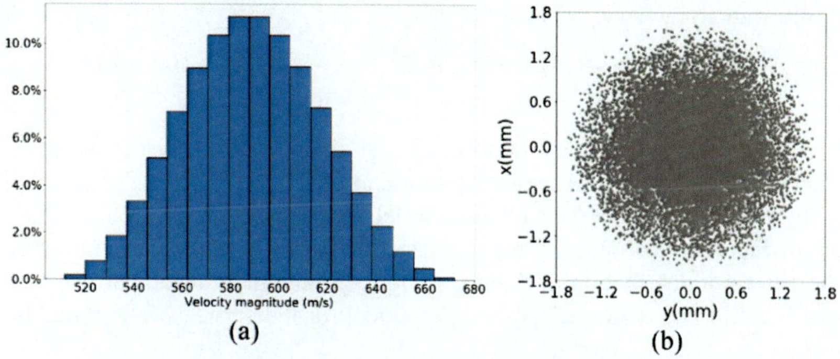

Fig. 2. Probability of the particle velocity (a) and spatial distribution (b).

This means that the COR has quite minor effect on the particle trajectories and acceleration for cold spray application. This finding will simplify the CFD simulation of the cold spray process.

3.2 Effect of COR on Particle In-Flight Behavior in Shot Peening

The same strategy was also implemented to the shot peening process simulation. The computed mean particle velocity and spatial distribution are tabulated in Table 1.

The mean particle velocity differs between cases, with difference as large as ~20 m/s, if the particles are injected from the injection port, which is the condition as in the reality. Therefore, COR plays a prominent role in the particles fate, as expected. What causes the difference of the effect of COR on cold spray and shot peening? The most possible reason is that the injection port of cold spray is very close to the nozzle, while that of shot peening is very far away from the nozzle. In such case, powders experience much more times of rebounding in shot peening than cold spray.

The simulation series of cases with the injection location moved to the nozzle inlet are intentionally conducted to prove this conjecture. As shown in the 4th row of Table 1, the computed mean velocity becomes very close to each other for all the 9 combinations of CORs. Therefore, we can draw a conclusion that the effect of rebounding is minor if the particle passage in the facility is as short as the same order of the nozzle.

The measured mean particle velocity from shot meter is 74.3 m/s, and some of the combinations of COR in the table have acceptable predicted value. However, the spatial

Table 1. Mean particle velocities (U_{mean}) and spatial distribution from simulation for different combination of CORs. Cases with the powders released from hose inlet and nozzle inlet are listed in row 3 and 4 respectively.

Case	1	2	3	4	5	6	7	8	9
Normal COR	1	0.8	0.8	0.8	0.8	0.8	0.7	0.5	0.4
Tangential COR	1	0.9	0.8	0.7	0.5	0.4	0.4	0.4	0.4
U_{mean} – hose inlet (m/s)	80.7	80.0	83.0	100.7	103.5	103.7	103.7	103.3	103.4
U_{mean} – nozzle inlet (m/s)	80.1	80.6	80.7	80.9	81.1	81.1	81.2	81.4	81.5
Particle distribution – hose inlet (mm)	36	9	9	12	11	12	10	11	38

distribution on the specimen is around 40 mm in diameter from visual inspection in experiments. This excludes all the cases with particles injected from the nozzle inlet, while only the case with perfect restitution delivers satisfied result. With consideration that perfect rebounding is not realistic, and the impact angle-dependent restitution formula delivers a distribution of diameter about 43 mm and a mean particle velocity 79.8 m/s, it could be considered as a proper model for the particle-wall rebounding in shot peening.

4 Conclusions

In this study, effect of the restitution of the particle-equipment wall on the particle trajectories and acceleration is investigated for the application of cold spray and shot peening. Considering the particle velocity and spatial distribution, coefficients of restitution (COR) play prominent role in the shot peening process while their role in cold spray is minor. This is attributed to the passage length of the process facilities, that is, shot peening facility has a very long hose from the injection port to the nozzle, while in cold spray the injection port is very close to the nozzle. For shot peening, angle-dependent coefficients of restitution, evaluated from the properties of the pair of impacting objects, are reasonable inputs to represent the particle-wall interaction.

Acknowledgements. Cold spray work is supported by Aerospace Program (1621500026), and the shot peening work is from an inter-RI project between IHPC and ARTC, ASTAR.

References

1. McCarty, J.L., Carden, H.D.: Impact Characteristics of Various Materials Obtained by an Acceleration-Time-History Technique Applicable to Evaluating Remote Targets. National Aeronautics and Space Administration, Washington, D.C. (1962)
2. Higham, J.E., Shepley, P., Shahnam, M.: Measuring the coefficient of restitution for all six degrees of freedom. Granul. Matter **21**(2), 1–5 (2019). https://doi.org/10.1007/s10035-019-0871-0

3. Dong, M., Mei, Y., Li, X., et al.: Experimental measurement of the normal coefficient of restitution of micro-particles impacting on plate surface in different humidity. Powder Technol. **335**, 250–257 (2018). https://doi.org/10.1016/j.powtec.2018.05.022
4. Ansys Fluent Theory Guide v19.0

Check for updates

Toolpath Planning for Cold Spray Additive Manufacturing

Hongjian Wu(✉), Mohamed Sokore, Wenbo Li, Rija-Nirina Raoelison, Hanlin Liao, and Sihao Deng

ICB UMR 6303, CNRS, Université Bourgogne Franche-Comté, UTBM, 90010 Belfort, France

Abstract. Cold spray (CS) has recently become one of the popular additive manufacturing (AM) processes, which allows forming large volume objects fast without the risk of undesirable heat effects. However, this unique technology still has some limitations in additive manufacturing application. Unlike other additive manufacturing technologies, CS is a continuous powder spraying and deposition process, which is more sensitive to spraying path precision and parameters. Therefore, common material filling paths are not suitable for cold spray additive manufacturing (CSAM). In this study, a tool path planning method for CSAM is proposed, which comprehensively considers the characteristics of CS, materials, spraying parameters and robot kinematics. Feature-based-layer built benchmark test objects have excellent shape accuracy. This implies that the proposed method makes CS a real and ready layer-by-layer AM process for 3D shape forming.

Keywords: Cold spray additive manufacturing · Tool path · Robot

1 Introduction

Nowadays, metal-related additive manufacturing (AM) technology has become the focus of the global industry. Cold spray (CS) [1] has attracted much attention because of its great potential in solid-state forming metal components. In fact, this technology now is regarded as a potentially competitive AM due to the ability to build 3D objects when the CS gun is attached to a robot. Nowadays, with the development of technology, it has gradually occupied its place in the field of AM. In comparison with other electron beams or laser radiations based AM technologies, many advantages make it unique in fabricating freestanding metal components.

In order to build 3D objects automatically, a space filling path planning is a necessity during AM. However, CS is a continuous powder spray and depositing method that is more sensitive to spraying path precision and parameters than other additive manufacturing methods. Many common methods for material filling paths planning are incompatible with cold spray additive manufacturing (CSAM). Therefore, this study aims to develop a tool path planning method for CSAM. This method takes into account the characteristics of CS deposition, materials, spraying parameters, and robot kinematics, enabling CSAM to elaborate freeform 3D objects with acceptable precision.

Y. Wei and S. Chng (Eds.): INCASE 2021, LNME, pp. 240–243, 2022.
https://doi.org/10.1007/978-981-16-5763-4_53

2 Spray Strategies for CSAM

As mentioned before, CS is a continuous powder spraying and depositing method. The distribution and velocity of particles in the feedstock jet are uneven [2], which causes the distribution of particles to nearly approximate Gaussian distribution [3]. In practice, an axisymmetric de-Laval nozzle is often used to emerge high-speed particles. When the nozzle traverse speed is low or the number of scanning passes is more, a triangular coating profile will be formed and the subsequent deposition of particles will be difficult. This characteristic would limit the application of CSAM for shape forming if no spray strategy for processing adaptation is used. For that purpose, Pattison et al. [4] proposed a strategy of triangular-tessellation. The process is to tilt the nozzle by a 4/5-axis system so that the current coating is sprayed perpendicularly onto the inclined surface of the previous track. In our previous work [5], a stable layer-building strategy was proposed for CSAM. There, three key parameters controlled by the robot were used to adjust the coating morphology in order to avoid forming triangular-like deposits. The inclined spray manner in both methods is to compensate for the thickness difference between the middle and side of the deposit.

It should be noted that the implementation strategy of CSAM should be adjusted according to the spraying system, which involves spraying equipment, materials and spraying parameters. The size of the nozzle outlet determines the distance between the two scanning tracks during the material filling process, which effects the filling density and the thickness of layer. In addition, the powder feed rate as well as the scanning speed also greatly affect the thickness of the layer, while he toolpath's upward movement is dependent on the thickness of the layer. The thickness of one layer can be obtained by a single pre-spray. Moreover, the filling process of each layer needs to consider the drop in shape and dimensional accuracy caused by the thickness loss of the edge. The degree of this edge loss is linked to the material sprayed. As shown in Fig. 1, the edge loss of different materials is different. This can be associated with the material characteristic and the deposition efficiency. Thus, the edge thickness compensation needs to be performed according to the actual situation, or it can be removed as an extra part.

Fig. 1. (a) Al deposits, (b) Cu deposits and (c) steel deposits by CSAM.

3 Tool Path Planning for CSAM

Since 3D forming via CSAM is no longer a process of producing simple coatings, tool path for space filling is the most critical. Here, a stable layer-by-layer building strategy and capability of CSAM is a key preliminary step for forming complex 3D shapes in an additive way. This can be accomplished in two steps: first, the surface of the prescribed area is sprayed in a perpendicular manner by specifying a scanning step of the corrected distribution. Here, the grating pattern can be used to determine the filling path in the area, such as the three typical path patterns shown in Fig. 2. Then, the edges are sprayed by tilting the nozzle to compensate for the edge loss, as described in prior work [5]. Afterward, a 3D volume is built in multiple layers by moving the tool path upwards. The upwards value is a layer thickness. Many other 3D printing technologies, such as fused deposition modeling (FDM), selective laser sintering (SLS), and selective laser melting (SLM), can stop feeding or incorporate 'skip' movements where no material is printed, whereas CSAM is a continuous powder spraying process, and switching between quick stop and quick start feeding is difficult due to the long response time of spraying system. In order to avoid the loss of shape accuracy caused by the intersecting continuous paths when filling the entire 3D volume, the following useful strategies can be used during tool path planning:

1) An end-to-end connection method can be adapted for tool path loops. As shown in the Fig. 3, the ending point of n layer connects to the starting point of n + 1 layer.
2) Irregular areas can be divided into several areas for filling with materials. The organization of the tool path needs to avoid too many turns and transformation, which will affect the robot performance and the deposition accuracy.
3) A slower nozzle traverse speed is used for material filling, while a higher nozzle traverse speed is applied to move swiftly from one region where material is printed to another. Here, a large speed difference can reduce the influence of overspray and unnecessary material filling on the shape accuracy.
4) Path moving out of the spraying area or contour path around the outsides of the part are allowed to avoid 'skip' movements.

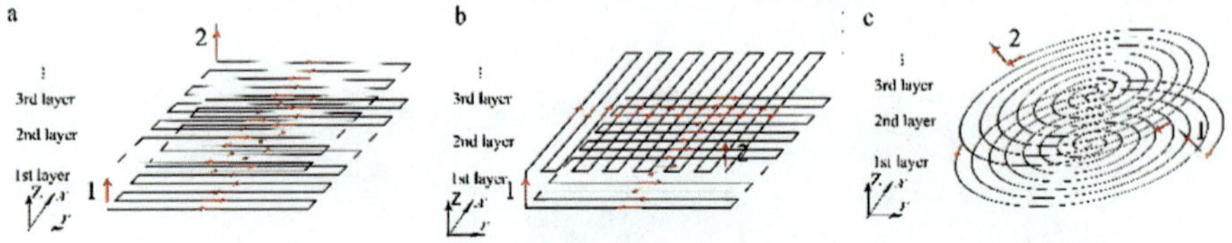

Fig. 2. Layer-by-layer filling with (a) zigzag path, (b) cross path, (c) spiral path.

In this study, a self-developed software is used to generate a continuous and non-intersection tool path for CSAM. The tool path can be exported to the robot language run by ABB robot. Meanwhile, the robot offline programming system, RobotStudio™, is used for CSAM process simulation. Through its trajectory simulation module, it can be observed whether there is divergence in between the expected and the actual robot trajectories. The analyser module of RobotStudio™ can also be used to record and analyse the position of each axis as well as the real TCP speed, so as to re-adjust the configuration of each axis. Once the trajectory of the robot is confirmed, the corresponding program can be downloaded to the actual robot system for calibration, testing and application. As shown in Fig. 3, some benchmarking cases with basic objects were carried out with the assistance of RobotStudio™. The results had great shape accuracy, which demonstrates that CSAM applications can benefit from the suggested tool path planning and generation method.

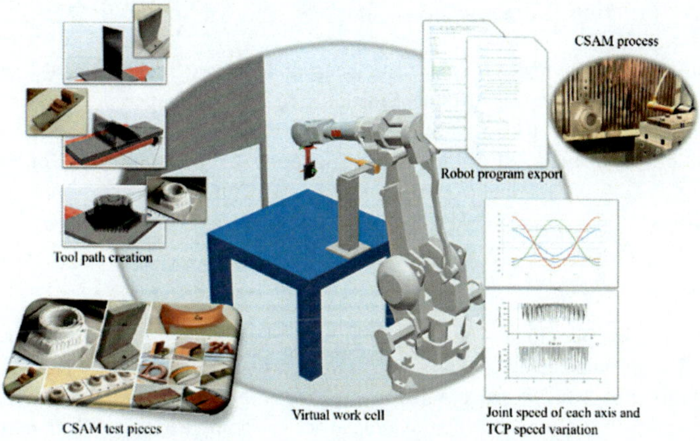

Fig. 3. Some basic objects produced by CSAM.

4 Conclusions

CSAM enables the batch fabrication of large and simple 3D volumes, as well as the small-scale manufacturing of custom-made items. A feasible tool path plays a decisive role in this process. The proposed tool path planning method and manufacturing strategies in this research realize the CSAM with high efficiency and high shape accuracy. New developments of CSAM may rely on increased spatial resolution by using micro-nozzles or other structural nozzles to achieve higher dimensional precision in 3D shape forming.

References

1. Raoelison, R., Verdy, C., Liao, H.: Cold gas dynamic spray additive manufacturing today: deposit possibilities, technological solutions and viable applications. Mater. Des. **133**, 266–287 (2017)
2. Gärtner, F., Stoltenhoff, T., Schmidt, T., Kreye, H.: The cold spray process and its potential for industrial applications. J. Therm. Spray Technol. **15**(2), 223–232 (2006)
3. Wu, H., et al.: A new approach to simulate coating thickness in cold spray. Surf. Coat. Technol. **382**, 125–151 (2020)
4. Pattison, J., Celotto, S., Morgan, R., Bray, M., O'neill, W.: Cold gas dynamic manufacturing: a non-thermal approach to freeform fabrication. Int. J. Mach. Tools Manuf. **47**(3–4), 627–634 (2007)
5. Wu, H., et al.: Stable layer-building strategy to enhance cold-spray-based additive manufacturing. Addit. Manuf. **35**, 101–356 (2020)

Check for updates

Computational Fluid Dynamics Simulation of Cold Spray Process for Particle Velocity Investigation

Aravinth Kamaraj[1], Wei Zhou[1], and Xu Song[2](✉)

[1] School of Mechanical and Aerospace Engineering, Nanyang Technological University, 50 Nanyang Avenue, Singapore 639798, Singapore
aravinth002@e.ntu.edu.sg, mwzhou@ntu.edu.sg
[2] Department of Mechanical and Automation Engineering, Chinese University of Hong Kong, Shatin, Hong Kong S.A.R., China
xsong@mae.cuhk.edu.hk

Abstract. Computational Fluid Dynamics (CFD) simulation is an effective method to optimize cold spray process parameters and coating quality. The objective of this work is to construct a CFD model to directly investigate influence of the process parameters such as pressure and carrier gas composition on the particles. This work describes a detailed study of the CFD approach and carrier gas composition, and how it affects the average particle velocity. A model of 3-dimensional nozzle with chamber and particle injector was used to generate the flow field of Ti6Al4V particles with help of compressed carrier gas such as nitrogen, helium at supersonic speeds. Particles are accelerated by the carrier gas to a very high velocity, resulting in plastic deformation upon impact with a solid substrate positioned at specific standoff distance. An optimum average particle velocity is found using CFD simulation for various compositions of carrier gas for a given particle size distribution.

Keywords: Cold spray · Computational Fluid Dynamics · Particle velocity

1 Introduction

Cold spray is gaining attention in recent years, due to its relatively low temperature and cost comparing to other types of spray techniques. Many institutes have setup and developed cold spray process for various alloys [2] and have achieved coating of desirable mechanical properties. However, the interfacial mechanical property is still lower than that of the coating or the substrate material, making it unsuitable for some load bearing applications. Moreover, the coating interface behaviour is closely linked with the choice of processing parameters. Therefore, it is our aim here to use the Computational Fluid Dynamics (CFD) simulation to understand the influence of the direct process parameters (i.e., pressure and temperature of the gas) on the cold spray and find an optimal particle velocity for bonding through the means of correlating the process parameters with the

Y. Wei and S. Chng (Eds.): INCASE 2021, LNME, pp. 244–248, 2022.
https://doi.org/10.1007/978-981-16-5763-4_54

particle velocity distribution. Although CFD simulation model has been proposed previously for cold spray [1], it is not yet well understood numerically, especially for the gas mixing scenario, which will further accelerate the particles and enhance the mechanical properties at the interface.

2 Numerical Method

The flow is considered as single-phase flow as He and N_2 can be well mixed for the cold spray application [3], and the dispersed phase model is used for particle modelling. Therefore, the discrete phase model - a Eulerian-Lagrangian approach, is employed in ANSYS® Fluent for multiphase modelling. In which, the continuous phase will be solved by using the Navier-Stokes equation. Similarly, the dispersed phase trajectories will be solved by tracking a large number of particles using equation of motions i.e. Lagrangian framework. As the dispersed particles only occupy a low volume fraction (normally <2%), although the mass of the particles is greater than the mass of fluid ($m_{particle} > m_{fluid}$), it is still tolerable to assume that the exchange of momentum and energy between the two phases only goes one-way, meaning only to consider the drag and lift of the gas exerting its effect to the particles. The flow is considered to be 3D compressible with species transport involving N_2 and He. Mesh independent study has been done with medium and coarse meshes before conducting simulation with fine mesh. The energy equation should be turned on, and for the turbulent modelling, Shear Stress Transport (SST) k-ω and k-ε models are employed.

Figure 1 illustrates the 2D cross-section and 3D model where we can see the boundary conditions. The combination of hexahedral and tetrahedral elements is used for meshing

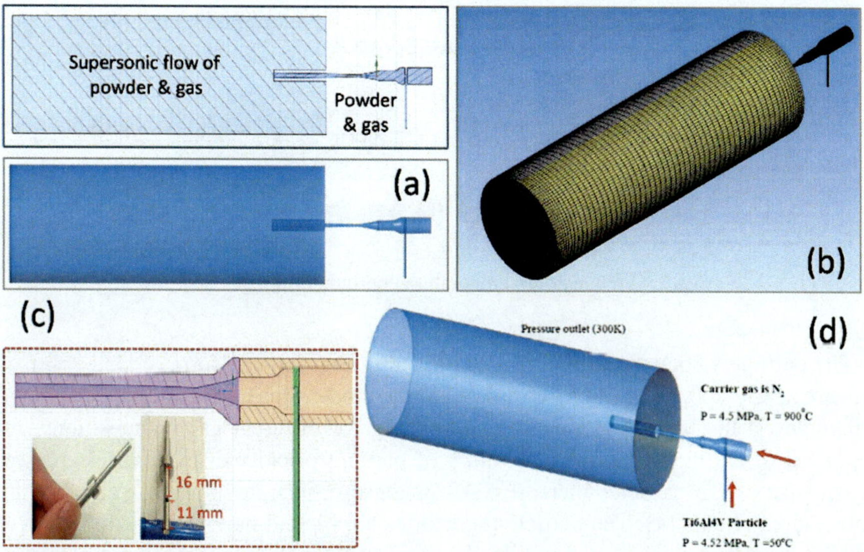

Fig. 1. Computational domain – illustration of (a) 2D cross-section; (b) 3D mesh; (c) details of the nozzle; (d) boundary condition.

the domain. Inside the core, a fine mesh is used, and in the outer domain, coarse mesh is used. An unstructured mesh is also employed to bridge the two. The total element count is 5.84 million. There are 3 main areas for boundary conditions, such as gas inlet, powder inlet and gas flow outlet. The carrier gas used is combination of N_2 and He in different compositions such as 50% N_2 and 50% He. The feedstock used is Grade 5 Ti6Al4V powders from GE Additive of sizes ranging from 15 μm to 45 μm and the mean diameter of 36 μm.

3 Results and Discussion

The velocity and temperature contours on the central plane are shown in Fig. 2(a)(b) and 2(c)(d) respectively. It can be observed that the flow velocity tends to increase when the proportion of helium used in the gas mixture is increased. For both flow operating conditions, the supersonic flow is produced, which is marked by the formation of diamond shocks downstream of the nozzle when the flow undergoes expansion. From the temperature contours, we can observe that the core flow decreases with the increase in proportion of helium used in the mixture. Top section for the converging section of the nozzle is subjected to both cold and hot helium flow.

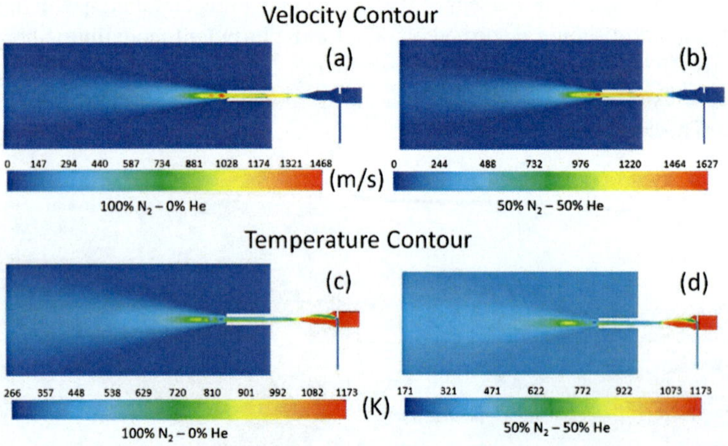

Fig. 2. Velocity and temperature contours of the gas flow

The particle velocity distributions are plotted in Fig. 3 in the form of histogram, as there are nearly 200,000 particles simulated in the model. It is noted that the percentage of particles is the highest at the average velocity for different gas compositions. As proportion of He increases, the uniformity of particle velocity distribution decreases, but velocity of the particles increases. All of these observations agree well with the experimental results [4]. The particle trajectories are plotted here for showing particle residence time history inside the nozzle. We notice that the design of the powder injector causes a sharp turn for the particle flow. Upon being injected into the stream, particles are pushed upwards because of momentum towards the outlet of the nozzle at high

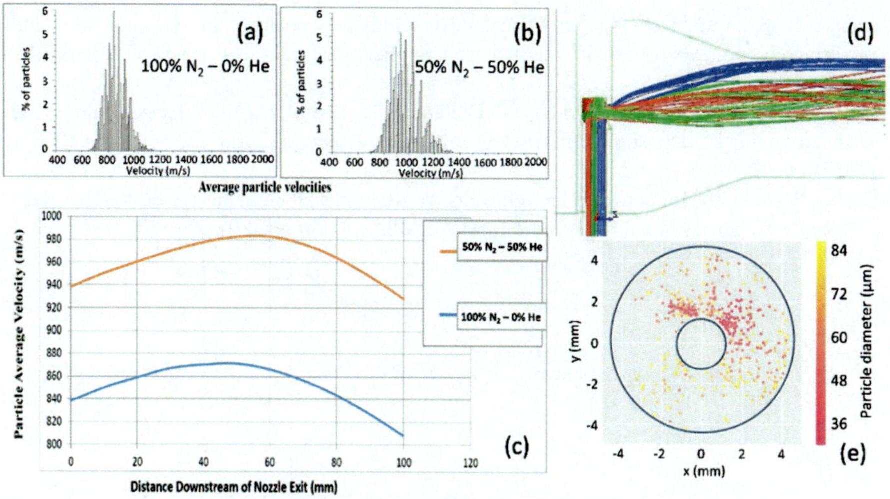

Fig. 3. Histogram of velocity versus percentage of particles for (a) 100% N_2 (b) 50% N_2 - 50% He and (c) average particle velocity versus nozzle exit distance (d) particle trajectory in the powder injector; (e) the powder injector inner wall cross-section particle distribution.

speed. We assume all the particles hit the inner wall of the nozzle are collected in the simulation. Then, a 2D cross-section of the nozzle bottleneck area is illustrated with the radial distribution of the particles. We notice that only very few particles are shown to hit the nozzle inner wall, and larger particles tend to distribute closer to the center. The possibilities to hit the nozzle inner wall is approximately less than 5%, which explains the experimental observation that clogging does not often occur during the cold spray process at these given gas conditions.

4 Conclusion and Future Works

The cold spray process parameters are shown by the CFD study to have direct influences on the particle velocity, which can be used as input for the following finite element simulation of particle impact process to assist the further investigation on the interfacial bonding strength. Further investigation of using Reynolds Stress Model to study the bow shock effect and how it will affect the particle flow in the supersonic regime adds good value to the current CFD study. The particle-particle interaction before impact is also a topic of great interest. These findings may facilitate further breakthrough in coupled Eulerian-Lagrangian CFD simulation of MCS process, which is an ongoing investigation that inspires the research community.

References

1. Karimi, M., Fartaj, A., Rankin, G., Vanderzwet, D., Birtch, W., Villafuerte, J.: Numerical simulation of the cold gas dynamic spray process. J. Therm. Spray Technol. **15**(4), 518–523 (2006)

2. Pathak, S., Saha, G.: Development of sustainable cold spray coatings and 3D additive manufacturing components for repair/manufacturing applications: a critical review. Coatings **7**(8), 122 (2017)
3. Ozdemir, O., Widener, C., Helfritch, D., Delfanian, F.: Estimating the effect of helium and nitrogen mixing on deposition efficiency in cold spray. J. Therm. Spray Technol. **25**, 660–671 (2016)
4. Tan, A.W.Y., et al.: Influence of particle velocity when propelled using N_2 or N_2-He mixed gas on the properties of cold-sprayed Ti6Al4V coatings. Coatings **8**(9), 327 (2018)

Analysis of Glass Transition Temperature Shifts Induced by Internal and External Plasticizers in (Meth)acrylate Polymer Coatings with Molecular Simulations

J. M. Phang[1], Satyasankar Jana[2], Alexander van Herk[2], and M. Klähn[1](✉)

[1] Institute of High Performance Computing, A*STAR, 1 Fusionopolis Way, #16-16 Connexis, Singapore 138632, Singapore
marco_klaehn@ihpc.a-star.edu.sg
[2] Institute of Chemical and Engineering Sciences, 1 Pesek Road, Jurong Island, Singapore 627833, Singapore

Abstract. Molecular dynamics (MD) simulations in combination with molecular mechanical (MM) force fields are used to quantify changes in glass transition temperature, T_g, of (meth)acrylate copolymers, induced by addition of plasticizers. Free external plasticizers, such as ethylene glycol-2-ethylhexyl ether, 1-phenoxy-2-propanol and butyl carbitol are observed to lower T_g by about 50 K upon addition to the polymer. After integration of these plasticizers into the copolymer through covalent binding, the T_g-loss caused by external plasticization partially recovers. The extent of observed T_g-loss recovery is found between 10 and 24 K and depends on structural details of the used external plasticizer. It is observed that compact external plasticizers undergo more frequent diffusive motion inside the polymer matrix due to reduced steric hindrance, thereby causing a larger reduction of T_g. Hence, binding of such a free compact plasticizer also leads to a larger recovery of the initial T_g-loss compared to a bulkier plasticizer. An understanding of the relation between plasticizer structure and polymer T_g can be used to control T_g-loss recovery upon binding of a free reactive plasticizer to the polymer [1]. The aim is to design suitable polymer – plasticizer coatings, in which such reactive external plasticizers covalently bind to the polymer, after applying the coating, to avoid leaking of potentially harmful plasticizers into the environment. At the same time, a sufficiently high T_g of the polymer can be ensured after the application of the coating and binding of the plasticizer [2, 3].

Keywords: Molecular simulations · Polymer coatings · Glass transition temperature · Plasticizer · Molecular dynamics

1 Background

The use of plasticizers in coatings is necessary to achieve a sufficiently low viscosity for application of the coating material. Such small organic external plasticizers are prone to leak into the environment with potentially detrimental ecological impact.

Y. Wei and S. Chng (Eds.): INCASE 2021, LNME, pp. 249–252, 2022.
https://doi.org/10.1007/978-981-16-5763-4_55

Hence, it would be desirable to bind reactive free plasticizers to the coating polymer after application to prevent their evaporation. Turning free plasticizers into internal plasticizers is accompanied by a change in T_g of the polymer. Understanding the changes in T_g induced by plasticizers before and after integration into the polymer is important to choose suitable plasticizers for an intended application.

2 Results

A multi-step MD simulation protocol was employed to thermally equilibrate copolymer systems including internal and external plasticizers. An equilibrated structure of the copolymer is shown in Fig. 1a. A simulation of these systems at different temperatures enabled determination of T_g by plotting the system volume, i.e. the reciprocal system mass density, as a function of temperature (Fig. 1b).

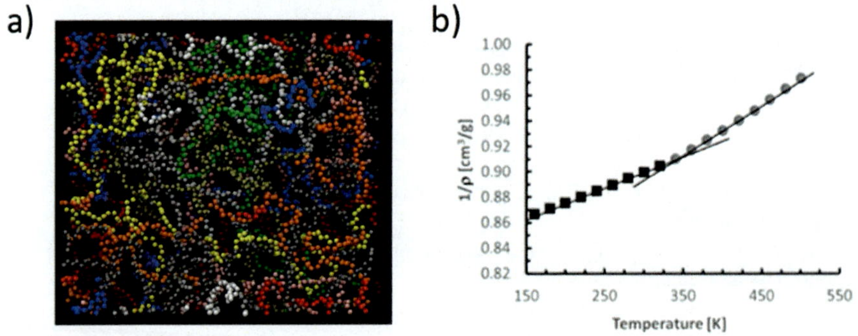

Fig. 1. a) Representative equilibrated atomistic structure of (meth)acrylate copolymer. Polymer chains are shown in different colors. b) Glass transition temperatures are determined with a volumetric method that derives the simulation volume as a function of temperature, V(T). A change in slope of V(T) indicates T_g, which was confirmed further by finding a maximum of the second derivative of V(T) at the same temperature.

By comparing T_g-values in the polymer system without plasticizer and with internal or external plasticizer, induced T_g changes could be derived as shown in Fig. 2. A comparison of structural features of the plasticizer together with an analysis of its diffusive behavior in simulations enabled us to relate plasticizer structure with T_g-loss recovery upon integrating the external plasticizer into the copolymer, as described in Fig. 3.

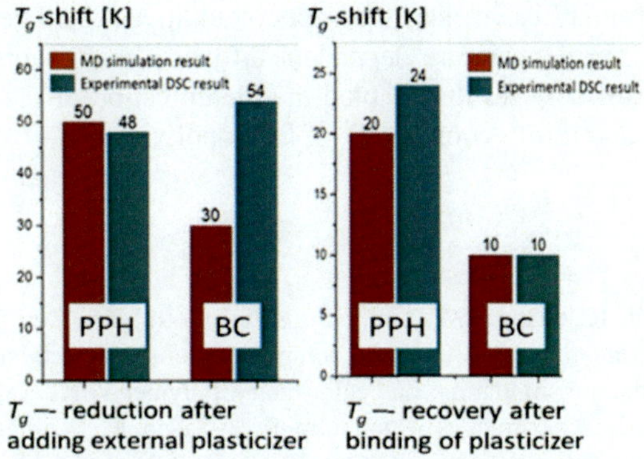

Fig. 2. Simulated T_g reduction of the copolymer after addition of external plasticizers (left side) in comparison with measured T_g losses for 1-phenoxy-2-propanol (PPH) and butyl carbitol (BC). T_g-loss recovery, as derived from simulations and compared to measurements, is shown on the right side.

a)

Why plasticizers reduce T_g:

External plasticizer	Internal plasticizer
(1) Reduced polymer entanglement and increased average inter-chain distance	(1) Reduced polymer entanglement and increased averaged inter-chain distance.
(2) Free movement of small plasticizers stimulates polymer movement	Plasticizer mobility restricted, i.e. no contribution to T_g-loss → T_g - loss recovery

b)

PPH	BC

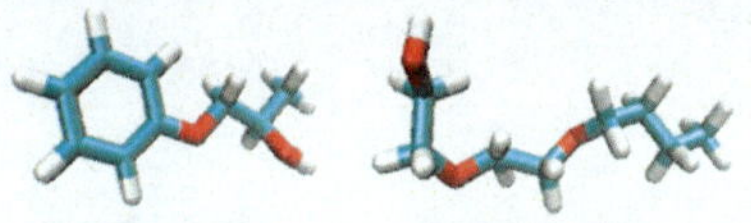

PPH	BC
• More compact • Less increase in inter-polymer chain distance	• Less compact • More increase in inter-polymer chain distance
• Frequent diffusive movement	• Less diffusive movement because bulkier
Effective external plasticizer (-48 K)	Effective external plasticizer (-54 K)
Less effective internal plasticizer (-24 K)	More effective internal plasticizer (-44 K)
Larger T_g-loss recovery (+24 K)	**Smaller T_g-loss recovery (+10 K)**

Fig. 3. a) Factors that cause reduction of T_g of a polymer when external and internal plasticizers are added, respectively. b) A comparison of two plasticizers, PPH and BC, regarding their ability to effect T_g-values of a polymer.

3 Conclusion

We developed a computational approach towards derivation of accurate T_g-shifts induced by internal and external plasticizers. Moreover, aggregation behaviour of additives inside the polymer matrix was analysed, highlighting some of the heterogeneous properties of the system. In addition, diffusive energy barriers of additives could be determined from simulations by deriving temperature dependent diffusion coefficients of the plasticizers. A combination of these results enabled the identification of the relation between plasticizer structure and its effect on the T_g of the copolymer.

References

1. Klähn, M., Krishnan, R., Phang, J.M., Lim, F., van Herk, A.M., Jana, S.: Effect of external and internal plasticization on the glass transition temperature of (Meth)acrylate polymers studied with molecular dynamics simulations and calorimetry. Polymer **179**(17), 121635 (2019)
2. Lahtinen, M., Glad, E., Koskimies, S., Sundholm, F., Rissanen, K.: Synthesis of novel reactive coalescing agents and their application in a latex coating. J. Appl. Polym. Sci. **87**, 610 (2003)
3. Barbosa, J.V., Veludo, E., Moniz, J., Magalhaes, F.D., Bastos, M.M.S.M.: Synthesis and characterization of acrylic fatty acid derivative and use as reactive coalescing agent. Eur. J. Lipid Sci. Technol. **114**, 1175 (2012)

Automatic Mirror Polishing of Rough, Freeform AM Surfaces Using a Toolpath Generation Framework

Josh Chen Ye Seng[1], Bisma Mutiargo[1](✉), Srinivasan Lakshminarayanan[2], Zhen Xie[1], Lim Pei Xian[1], Mohamed Faizal Hussain[1], and Tan Sui Wei[1]

[1] ARTC, A*STAR, Singapore, Singapore
{josh_chen,mutiargob,xie_zhen,limpx,mohamed_faizal_hussain, tansuw}@artc.a-star.edu.sg

[2] School of Mechanical and Aerospace Engineering, NTU Singapore, Singapore, Singapore
srini.gln@ntu.edu.sg

Abstract. Despite the emerging adoption of robots in the manufacturing industry, the mirror polishing processes are still dependent on skilled labour. As a labour-intensive process, mirror polishing requires an extensive number of man-hours to obtain high-quality surface finishing. Traditional automation techniques that rely on robot programming are extremely tedious and time-consuming. To automate the mirror polishing process and eliminate the bottlenecks associated with traditional robot programming, we propose a simple yet robust mirror-polishing framework. In our framework, a simple graphical interface processes the 3D point cloud and generates the toolpath using the user selection. Subsequently, the real robot automatically executes the generated toolpath through ROS. An intuitive graphical interface was used to reduce the requirement of skilled operators for the process. Stainless steel (MS1) work pieces of varying geometries were used in the trials. In preliminary validation results, our framework achieved a high-quality mirror surface finishing with desired geometric profile and roughness value of Ra 0.5 μm with a distribution of 0.1 μm (1 σ) .

Keywords: Robot toolpath generation · Robotic polishing · Mirror polishing

1 Introduction

As the final stage of the manufacturing pipeline, polishing is an extremely important process that improves the final quality, aesthetics and extends the lifespan of the component [1]. Due to its complexity, the mirror polishing process is predominantly carried out by highly skilled labour [2], leading to long manufacturing cycle time, variabilities in output, need for reworks, etc. Hence, there is a great demand for automating the mirror polishing process and obtaining high-quality finishing. While the repeatable and low-skilled tasks are easier to automate, specialized tasks such as robotic polishing require skilled labour for programming the robot, which is a bottleneck in the adoption of robots. This difficulty compounds in a high mix - low volume production setup that is becoming

Y. Wei and S. Chng (Eds.): INCASE 2021, LNME, pp. 253–257, 2022.
https://doi.org/10.1007/978-981-16-5763-4_56

more common, especially in the Small-Medium Enterprises (SME). Simplifying robot programming is an open research problem and- an active area of research [3–6]. In this paper, we present a simple polishing framework that can obtain high-quality surface finishing through an intuitive graphical user interface (GUI). The GUI that we developed can obtain the point cloud from the 3D scanner or CAD model, and automatically carry out semantic segmentation to cluster the part into different surfaces. The operator can select various surfaces in the GUI and generate the desired toolpath in real-time. Finally, the user can generate the robotic toolpath using the GUI and execute the robotic motion. The main contribution of this paper is our intuitive framework that can accomplish high-quality mirror polishing. The paper is organized as follows: Sect. 2 provides the methodologies of the framework; Sect. 3 presents the hardware setup and the trials; Sect. 4 shows results and discusses key findings based on the results.

2 Methodology

The proposed framework comprises data acquisition using Artec Eva 3D scanner, pre-processing, surface identification and toolpath generation, and finally, robotic execution as shown in Fig. 1. The point cloud data of the component geometry, along with the work environment was first acquired using the 3D scanner. The data is pre-processed to remove noises and background information. The coordinate system of each point is transformed to align with the robot's tool coordinate point (TCP) by using 3D feature detectors. Once the pre-processing completes, the point cloud is segmented to identify the different surfaces in the 3D environment using the PCL library. Users can subsequently select the required surfaces and toolpath surface patterns such as Zigzag, Spiral etc., intuitively by interfacing with the GUI, as shown in Fig. 2 below.

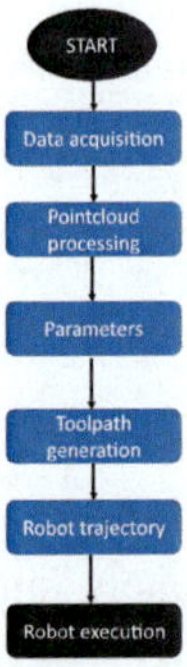

Fig. 1. System process flow

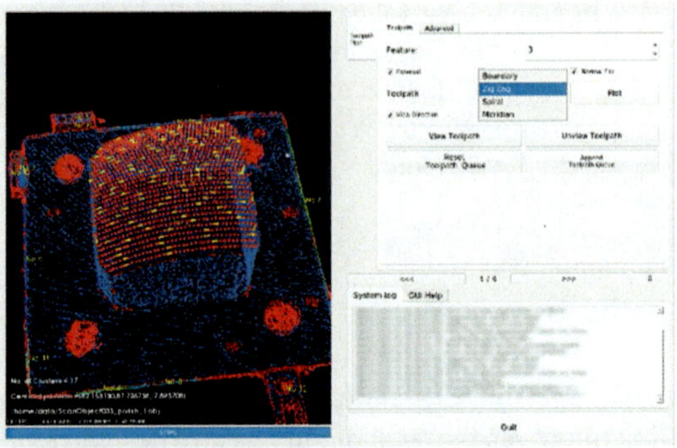

Fig. 2. Toolpath generation GUI

Once the desired toolpath is generated, it is projected onto the 3D Cartesian system with normal directions on the component surface. Using the motion planner in ROS, the software computes a via robot motion for the desired toolpath accounting for collision and pose viability. In addition, the operator can select multiple process parameters such as force, spindle speed, tool path recommender, feed rate, and the end-of-arm tools in the

GUI. Thus, through the GUI, even unskilled labour can accomplish the entire toolpath generation pipeline from pre-processing to the robot execution without any knowledge of robots or programming.

3 Experimental Setup and Trials

The hardware setup consists of ABB 1200 robot, fixtures for fastening the work-coupons, ATI Delta F/T sensor, NSK Nakanishi spindle, as shown in Fig. 3 below. The ABB 1200, a Six degree of freedom (DoF) industrial robot, is selected for the polishing trials due to its high accuracy. We have utilized an ATI Delta IP65 Multi-Axis Force/Torque Sensor system that measures all six components of force (Fx, Fy and Fz) and torques (Tx, Ty and Tz).

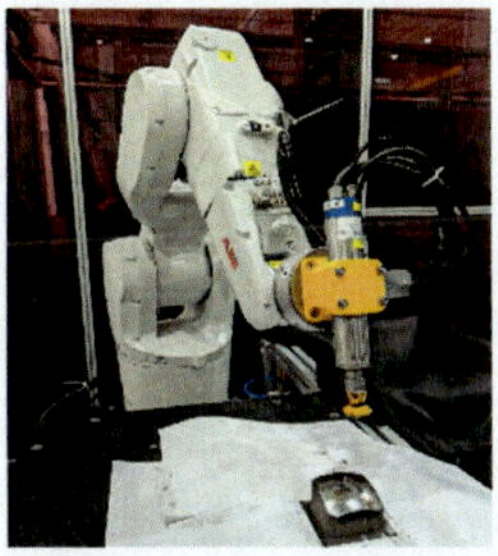

Fig. 3. Hardware setup

Fig. 4. Printed wavy and dome coupons

Trials on curved stainless steel (MS1) coupons were carried out to validate the capability of the framework. The coupons dimension are 50 × 50 mm of 2 surface geometries, wavy and dome design as shown in Fig. 4 above. They are printed on an EOS powder bed machine, and cut using electrical discharge machining (EDM) wire cutting. The average initial roughness value is Ra 18 μm with 6.5 μm variance (1 σ) for the dome design, and 23 μm with 7.9 μm variance (1 σ) for the wavy design. Mirror polish can only be obtained in multiple iterations by using abrasives with a gradual increase of the grit size. In total, we conducted a set of trials by varying the key process variables (KPVs) such as Force, Spindle speed and Feed rate with the variation as shown in Table 1 below.

4 Results and Discussions

Visual qualitative analysis indicates a promising result with the wavy and dome design. Figure 5 shows the appearance of the wavy and dome coupons before polishing and after polishing, indicating that the mirror polishing can be achieved automatically with our toolpath generation program. Preliminary surface roughness analysis conducted with white light interferometry done on the dome coupon shows that the roughness has decreased from 18.68 μm, with 6.56 μm standard deviation throughout six different points on the surface, to 0.56 μm with 0.14 μm distribution throughout the entire surface. The preliminary result shows that there is a great promise to use our automated toolpath generation program to do the polishing process on raw additive manufacturing surfaces.

(a) (b) (c) (d)

Fig. 5. Before polishing of wavy and dome coupons (a,b); after mirror polish inishing of wavy and dome coupons (c,d)

Table 1. Details of the trials

Abrasive	Feedrate (mm/s)	Force (N)	Spindle Speed (rpm)	Iterations (#)
Grit 60	2	3	5000	4–14
Grit 150	2	5	5000	4
Mesh 200	2	10	6000	8
Mesh 1200	2	5	8000	4
Mesh 8000	2	5	8000	8–12

5 Conclusion

In this paper, we presented a novel toolpath generation framework for mirror polishing to mitigate robot programming. The proposed framework is highly suitable for SMEs that have low-volume-high-mix product catalogue and cannot afford the long labour time and high labour cost associated with manual programming. Our preliminary trials on stainless steel coupons indicates that the methodology is able to bring down the surface roughness of the dome coupon from 18.68 μm with 6.5 μm variance (1 σ) throughout, to a mirror finish, 0.56 μm with 0.14 μm distribution (1 σ) throughout the surface.

References

1. Kana, S., Lakshminarayanan, S., Mohan, D.M., Campolo, D.: Impedance controlled human–robot collaborative tooling for edge chamfering and polishing applications. Robot. Comput. Integr. Manuf. **72**, 102199 (2021)
2. Mohammad, A.E.K., Hong, J., Wang, D., Guan, Y.: Synergistic integrated design of an electrochemical mechanical polishing end-effector for robotic polishing applications. Robot. Comput. Integr. Manuf. **55**(February 2018), 65–75 (2019). https://doi.org/10.1016/j.rcim.2018.07.005
3. Zhen, X., Seng, J.C.Y., Somani, N.: Adaptive automatic robot tool path generation based on point cloud projection algorithm. In: 2019 24th IEEE International Conference on Emerging Technologies and Factory Automation (ETFA), pp. 341–347 (2019)
4. Lakshminarayanan, S., Kana, S., Mohan, D.M., Manyar, O.M., Then, D., Campolo, D.: An adaptive framework for robotic polishing based on impedance control. Int. J. Adv. Manuf. Technol. **112**(1–2), 401–417 (2020). https://doi.org/10.1007/s00170-020-06270-1

5. Pan, Z., Polden, J., Larkin, N., Van Duin, S., Norrish, J.: Recent progress on programming methods for industrial robots. Robot. Comput. Integr. Manuf. **28**(2), 87–94 (2012). https://doi.org/10.1016/j.rcim.2011.08.004
6. Lakshminarayanan, S., Manyar, O.M., Campolo, D.: Toolpath generation for robot filleting. In: Itoh, S., Shukla, S. (eds.) INCASE 2019. LNME, pp. 273–280. Springer, Singapore (2020). https://doi.org/10.1007/978-981-15-0054-1_28

Author Index

Y. Wei and S. Chng (Eds.): INCASE 2021, LNME, pp. 259–261, 2022.
https://doi.org/10.1007/978-981-16-5763-4